Saving the Dammed

Saving the Dammed

Why We Need Beaver-Modified Ecosystems

ELLEN WOHL

OXFORD
UNIVERSITY PRESS

OXFORD
UNIVERSITY PRESS

Oxford University Press is a department of the University of Oxford. It furthers the University's objective of excellence in research, scholarship, and education by publishing worldwide. Oxford is a registered trade mark of Oxford University Press in the UK and certain other countries.

Published in the United States of America by Oxford University Press
198 Madison Avenue, New York, NY 10016, United States of America.

CIP data is on file at the Library of Congress
ISBN 978–0–19–094352–3

1 3 5 7 9 8 6 4 2

Printed by Sheridan Books, Inc., United States of America

*To my mother, Annette Wohl, whose courage, compassion,
and cheerfulness continue to inspire me*

CONTENTS

The Beaver Meadow on North St. Vrain Creek

There is a place, about a mile long by a thousand feet wide, that lies in the heart of the Southern Rocky Mountains in Colorado. Here at the eastern margin of Rocky Mountain National Park, along a creek known as North St. Vrain, everything comes together to create a bead strung along the thread of the creek. The bead is a wider portion of the valley, a place where the rushing waters diffuse into a maze of channels and seep into the sediment flooring the valley. In summer the willows and river birch growing across the valley bottom glow a brighter hue of green among the darker conifers. In winter, subtle shades of orange and gold suffuse the bare willow stems protruding above the drifted snow. The bead holds a complex spatial mosaic composed of active stream channels; abandoned channels; newly built beaver dams bristling with gnawed-end pieces of wood; long-abandoned dams now covered with willows and grasses but still forming linear berms; ponds gradually filling with sediment in which sedges and rushes grow thickly; and narrow canals and holes hidden by tall grass: all of these reflect the activities of generations of beavers. This is a beaver meadow.

The bead of the beaver meadow is partly hidden, tucked into a fold in this landscape of conifers and mountains. The approach is from Route 7, which runs north–south across the undulating topography of creeks flowing east toward the plains. Coming from the north, as I commonly do, you turn west into the North St. Vrain watershed on an unpaved road perched on a dry terrace above the creek. The road appears to be on the valley bottom, but beyond the terrace the valley floor drops another 20 feet or so to the level at which the creek flows. I instinctively pause at this drop-off. The conifer forest on the terrace is open and the walking is easy. The beaver meadow looks impenetrable and nearly is. I have to stoop, wade, crawl, wind, and bend my way through it, insinuating my body among the densely growing willow stems and thigh-high grasses. Swift and easy movement is not possible, but slow and arduous can facilitate observation. As Mary Oliver wrote, "Attention is the beginning of devotion" (p. 8).

Any landscape can be viewed at close range or with the sweep of distance. This landscape invites the close view. The details revealed provide rich rewards, from an insect pagoda chewed and built from the stems of sedges, to the interlaced ice crystals lidding flickers of sunlight on flowing water. I enter this world and am quickly lost to view, leaving behind the trails and traffic of the nation's third-most-visited national park. I am in the beaver meadow and, if I think about it, in the watershed of North St. Vrain Creek in the Colorado Front Range in the Southern Rockies. Only secondarily am I in the constructs of human society—a national park or the state of Colorado. Primarily, I am immersed in a beaver world. From this immersion I emerge, like a beaver rising to the surface of a pond, to contemplate the wider world.

The creek is named for Ceran St. Vrain, a French fur trapper who came to the region in the early decades of the 1800s to trap beavers for their fur. St. Vrain was one of hundreds of fur trappers working their way across North America from the Eastern Seaboard to the Pacific Coast, systematically decimating the beaver populations that once spread from the Alaskan tundra down to northern Mexico.

Eurasian beavers (*Castor fiber*) were once abundant from eastern Siberia west to Scandinavia and from the coast of the Arctic Ocean south to Spain, Iraq and Iran, and the Korean Peninsula and China. The animals were hunted so intensively, however, that by 1640 they were scarce. By the early 20th century, only 1,200 beavers were estimated to remain in all of Eurasia. Long before that, the fur trade had moved to North America, severely reducing populations of North American beavers (*Castor canadensis*) between 1620 and 1850.

There is no evidence that St. Vrain or the other fur trappers in what became Colorado took any precautions to keep from killing all of the beavers in the region. On the contrary, when John Charles Frémont came through Colorado in 1842, he wrote of the numerous abandoned, derelict beaver lodges and meadows, and the absence of beavers. Beaver populations did eventually recover once fashions changed, reducing the demand for beaver fur to use in making gentlemen's hats.

Beaver numbers can grow rapidly, as evidenced by the beavers now present in Chile's Tierra del Fuego region, all of which descended from 25 pairs of beavers imported to the area in 1946. Although population estimates are not available, the beavers now occupy about 27,000 square miles of land along nearly 17,000 miles of rivers in the Chilean archipelago. But Tierra del Fuego is a remote region difficult to access and the beavers there are largely left alone. Although intensive trapping of beavers in North America declined after the early 19th century, the animals continued to be removed or displaced by people settling in as farmers, ranchers, miners, and city dwellers. Today, ecologists estimate that perhaps 12 million beavers live in North America, compared to as many as 400 million prior to the start of the European fur trade.

Despite the activities of St. Vrain and other trappers, a beaver meadow now occupies the 150 acres along North St. Vrain Creek that are the focal point of this book. The term "beaver meadow" is used by scientists to describe the wetlands created where beaver dams send water flowing over the streambanks and across the valley bottom. The term appears to have entered scientific writing in 1938, when Rudolf Ruedemann and W. J. Schoonmaker adopted it from common usage in Upstate New York. A much older version is preserved in the name Beverly, which derives from "beaver" and the Anglo-Saxon "lea" for meadow.

Lewis Henry Morgan wrote a particularly apt description of a beaver lea in his 1868 book *The American Beaver and His Works*:

> At a distance they appear to be level and smooth; but when you attempt to walk over them, they are found to be a series of hummocks formed of earth and a mass of coarse roots of grass rising about a foot high, while around each of them there is a narrow strip of bare and sunken ground . . . through which the water passes when the meadows are over-flowed. A beaver meadow, therefore, may be likened to the face of a waffle-iron—the raised eminences of which represent the hummocks of grass, and the indentations the depressions around them for the passage of water. (pp. 203–204)

One of the hidden passages of water in which the North St. Vrain beaver meadow is rich. A color version of this figure is included in the insert section.

The Great Drying

One of the most careful and persistent observers of beavers in Rocky Mountain National Park, or anywhere, was Enos Mills. Mills came to the region at the age of 14 and lived there until his death, teaching himself a great deal about local natural history through patient observation. The preface to his 1913 book *In Beaver World* describes the book as the product of nearly 27 years of observation. Mills lived only to 53, but he wrote numerous books over the course of his relatively short life. Subsequent scholarship suggests that he embellished some details of his adventures and observations, perhaps seeking to convey the spirit rather than the letter of his experiences, but most of the behaviors Mills ascribed to beavers have been supported by later scientific research, and his descriptions remain by far the most enjoyable to read. Mills anthropomorphized beavers in his descriptions and some of his choices of imagery seem quaint today. But the more people systematically study animals, the more clearly the evidence indicates that they are not all that different from humans. Many other species, especially mammals and birds, play, grieve, make and use tools, display distinctive personalities and learned behaviors, and employ some form of language to communicate. Mills observed all of these characteristics among the beavers of Rocky Mountain National Park.

The statement that figuratively stopped me in my tracks when reading his book was his description of nearly 40 beavers working within sight at one time. That number is now far greater than the total remaining beaver population in Rocky Mountain National Park. Mills also had the privilege of watching coyotes, mountain lions, and wolves prey on beavers. I can only dream of such sights, for wolves are extinct in the region and mountain lions are uncommon in the park.

Mills described beavers as the original conservationists, constructing dams and ponds that limit erosion, store sediment, and reduce flood damage. He recognized that beaver dams store water and help to sustain stream flow, a recognition that eluded the state engineer of Colorado nearly a century later. Mills wrote of how beaver ponds provide waterholes for fish and are helpful in maintaining deep waterways by reducing the extremes of both high and low water, as well as reducing the quantity of sediment carried into channels. And he understood that, despite the disappearance of the "millions of beaver ponds [that] graced America's wild gardens at the time the first settlers came" (p. 65), the beavers left a legacy of rich alluvial soils across the continent.

Mills's observations regarding the legacy left by beavers are particularly pertinent. As Eurasian and then North American beavers were hunted nearly to extinction, their dams fell into disrepair and disappeared. Floodwaters were more likely to remain within a single, main channel rather than dispersing into many secondary channels and spreading across the floodplain. Floods from melting snow and rain moved more quickly downstream rather than soaking into the floodplain soils, and the water table of the valley bottoms dropped. As the wet meadows and ponds of the beaver meadows dried, upland plants replaced the sedges, rushes, and willows along the valley bottoms.

Simultaneous with, or shortly after, the removal of beavers were the other changes in resource use that largely eliminated floodplain lakes and wetlands. Deforestation and direct removal of wood from streams and rivers eliminated the large log rafts and logjams that had forced flow over the channel banks in a manner similar to beaver dams. Channelization and the construction of levees increased the ability of main channels to convey flood waters downstream without overtopping their banks. Land drainage for agriculture further lowered the water table across many valley bottoms. These activities increased the size of the main channel, removed obstructions within the channel, and severed the connections between the channel and its floodplain. The cumulative effect was to dry up valley bottoms as diverse as small, headwater streams and the large, lowland portions of the Mississippi and other great rivers. Ecologists now estimate that more than 80% of the riverside marshes, swamps, lakes and ponds, and floodplain forests of North America and Europe have disappeared during the last 200 years.

The removal of beavers and other changes in vegetation and the configuration of the land surface created what I refer to as the Great Drying—a pervasive change in river networks that occurred over a period of centuries nearly everywhere in the temperate latitudes. Today, the few remaining large rivers in boreal and tropical regions that have not been extensively dammed, leveed, and otherwise altered—the Yukon, the Amazon, the Congo—give some hint of what rivers in the temperate zone once looked like. One of the most notable features of these rivers is their messiness. Walking across the valley bottoms of a big, natural river and its tributaries is extremely difficult. Numerous smaller channels branch from and rejoin the main channel. Downed trees lie thickly across the floor of the floodplain forest, decaying into soil and providing nursery sites for younger trees and habitat for birds, small mammals, insects, and fungi. Water, standing and flowing, is abundant. Lakes, swamps, and marshes supplied by subterranean and surface flows create a maze of wetlands among dense vegetation that varies from tall grasses or sedges to thickets of trees that thrive in the wet soils of these valley bottoms. Even traveling along these rivers by boat can be challenging where the channel splits into branches or nearly doubles back on itself as it winds through tight bends. Enormous trees float down the tropical rivers, and rafts of smaller trees block portions of the boreal rivers. Sandbars hidden by waters turbid with silt and clay form, erode, and reform elsewhere, and the current is as difficult to read as Mark Twain describes for 19th-century navigation on the Mississippi River. Floods are not dampened by dams, and the main river and its tributaries change from year to year. These rivers create a window onto the history of rivers in the temperate zone that have been long since simplified and regimented.

Intensive river engineering was not confined to the large rivers of the temperate zone. Historical accounts of rivers across the United States at the time of first European settlement are notable both for the changed world they describe and for the thoroughness with which people of European descent set about changing every portion of the river network, from the smallest headwater channels to the Mississippi and the Columbia Rivers. Historical records of logging from New York state and the Great Lakes region include descriptions of loggers floating cut timber down channels so small that a man could almost straddle the channel. If the channel was wider than the average tree diameter, it could be used for log floating, but only after the naturally occurring, downed wood was removed. In his book *Running the River*, Carlton Morrison quotes Georgia State Engineer Hamilton Fulton, who wrote of the Oconee River in 1827, "I have never seen a river so much obstructed by logs and snags" (p. 57), yet descriptions from rivers around the country contain similar comments. Jim Sedell and Karen Luchessa write of British Army journals describing the Pacific Northwest during the same period, which feature accounts of valleys so wet that trails followed the side slopes: the valleys were "mostly water connected by swamps" as a result of beaver activity. Describing an 1841 trip across the tallgrass prairie of Illinois, Eliza Steele wrote, "Here the road crossed the wet prairie . . . Through this wet land we went splash, splash, nearly half the night" (p. 123). Describing the clearing of a natural wood raft from Louisiana's Atchafalaya River during the period 1839 to 1860, historian Martin Reuss wrote in his 2004 book *Designing the Bayous* that "The Atchafalaya Basin supplied an inexhaustible conglomeration of stumps, limbs, branches, and natural debris to obstruct stream channels. Sometimes, the state engineers became desperate in their pleas for assistance, not only from the legislature, but from landowners along the waterways" (p. 35).

The use of rivers large and not so large for navigation and timber floating required that the rivers be "cleaned" of wood and beaver dams. Removal of these channel obstructions allowed peak flows to be contained within a main channel rather than spreading across the floodplain or into multiple channels, and valley bottoms across the country became drier. People wishing to settle on or farm the floodplains hastened the process by digging canals to drain the valley bottoms, as well as dredging the river channels and throwing up levees along their banks. People built dams to store the flood peaks and release the water more gradually throughout the growing season for crops, and they pumped the groundwater that supplied valley bottoms with springs and seeps. The net result of these changes was to dry and destroy the majority of river wetlands across the United States, from the small pools in which native fish of the shortgrass prairie took refuge during droughts to the extensive bayous created where enormous wood rafts forced water out of the main channel in lowland rivers of the southeastern United States.

This is the Great Drying of the past two centuries. We are only now recognizing some of the negative consequences, from alarming extinction rates of freshwater animals to widespread eutrophication of coastal regions such as the Gulf of Mexico as a result of excess nitrogen and phosphorus running off uplands across the country without the benefit of filtering by riverine wetlands. It is in this context that I seek to explore the microcosm of the beaver meadow on North St. Vrain Creek and draw from this exploration understanding that can inform our responses to the macrocosm of the Great Drying.

A Watery Microcosm

In Colorado, gold miners followed hard on the heels of the fur trappers, and farmers and loggers came in with the gold miners. Following the discovery of gold in 1859, the Colorado Front Range from the Wyoming border south to the divide between the South Platte and Arkansas River watersheds was nearly completely deforested by 1900. Along with massive timber harvest came floating of cut logs down rivers; construction of roads and railroads beside rivers in narrow mountain valleys; and diversion of stream flow for growing crops. Only a few watersheds remained relatively unaltered, one of which was North St. Vrain Creek between the Continental Divide and the base of the mountains. And only a few small stands of old-growth forest survived the tree cutting and the increase in forest fires. One of the largest of these surviving stands lies in the upper portion of the North St. Vrain Creek drainage. The North St. Vrain watershed is my window onto the history of rivers in the Colorado Rockies.

There is nothing particularly unique about the beaver meadow on North St. Vrain Creek; others like it enrich rivers throughout the mountains of North America. Yet this beaver meadow is the microcosm that reflects the macrocosm. The history of these 150 acres epitomizes 15,000 years of changes in the forested landscapes of Eurasia and North America, from the retreat of the great Pleistocene ice sheets through the development of agriculture and intensive human manipulation of Earth's surface. And the future of the beaver meadow on North St. Vrain Creek represents the future of life on Earth.

In this book, I use the microcosm of this one beaver meadow to explore in detail the implications of historical changes to rivers. Beavers still live in the North St. Vrain beaver meadow, allowing me to explore the consequences of their presence. During the past few decades, beavers have disappeared from former beaver meadows in the drainages adjacent to North St. Vrain Creek, so I can explore the consequences of their absence. Month by month in this book, I observe the seasonal changes in the North St. Vrain beaver meadow. From the platform of these observations I consider the larger context of beavers, river health, and landscape change across the Northern Hemisphere. William Blake wrote of seeing the world in a grain of sand; I see the world in the North St. Vrain beaver meadow.

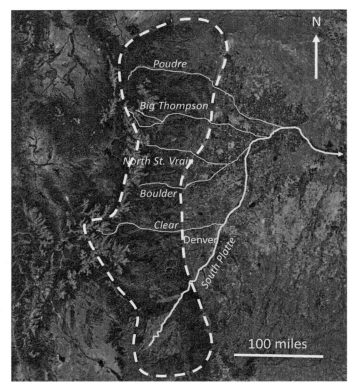

Location map of the Colorado Front Range (outlined in white) and the major tributaries of the South Platte River. The South Platte flows eastward to join the Platte River in Nebraska. A color version of this figure is included in the insert section.

A portion of the beaver meadow in early June, looking upstream toward the Continental Divide. A color version of this figure is included in the insert section.

The same pond, along the southern margin of the beaver meadow, covered with ice in November. A color version of this figure is included in the insert section.

What I see in this microcosm convinces me of the critical importance of protecting and restoring beaver meadows and beaver-modified landscapes across the Northern Hemisphere. The title of this book refers not to the beavers but to the environments that they create. I consider it imperative that we save the dammed— the beaver ponds and meadows that create so many vital functions within rivers— in order to protect stream flow, water quality, and abundant and diverse habitat for a wide variety of plants and animals.

January

Of Rocks and Ice

The beaver meadow is quiet in January. For many plants and animals, winter is a season of subdued activity, or of waiting. North St. Vrain Creek remains open along the main channel, the water flowing clear but tinted brown as pine bark between snowy banks. Densely growing thickets of willow closely line the banks. Each stem starts pale brown near the ground, then grades upward to shades of maroon or yellowish orange at the branch tips. In a bird's-eye view, these startling colors make the meadow stand out distinctly from the dark green conifers that define the edges of the meadow. Spruce and fir trees grow sharply pointed as arrows; pines present a slightly more rounded outline. Snow falls silently in thick flakes from the low, gray sky. The upper edges of the valley walls fade into snow and clouds. The sun appears briefly as a small, pale spotlight behind the clouds to the south.

Snow mounds on the patches of ice in the shallow channel. The water flowing beneath creates flickers through the translucent ice like a winter fire of subdued colors and no heat. Tussocks form humps of straw-colored grass above the dark, frozen soil. Rabbit tracks line the snowy bank, sets of four paw marks with a large gap between each set. Something small crossed the bank, leaping one to two feet at a bound, two paws with slight drag marks behind them. In places the powdery snow has drifted deeply, but mostly it is shallow over a frozen crust. Beaver-gnawed sticks and stumps poke up through the snow. A large flood came through four months ago, in mid-September, washing out dams that the beavers have not yet rebuilt. Chunks of wood deposited among the willow stems by the floodwaters stand far above the January flow of the creek.

A dipper fishes the creek, wading rather than swimming, at home in the cold water. The slate-gray bird is the only visible animal, busily probing the bed with its short bill, then pausing to stand and bob up and down. The dipper eats well on larvae of caddisflies and mayflies nestled among the streambed cobbles. The water is so shallow that the bird's short, thick body is only partly submerged. A second dipper stands on an emergent cobble, as though trying to warm its toes, then jumps

Looking upstream along a side channel of North St. Vrain Creek. The willows along the valley bottom give way to conifers on the valley side slopes. A color version of this figure is included in the insert section.

to a snowbank and lands in the stream with a plop, body surfing below the water and emerging beside the first dipper to stretch its neck upward and sing briefly. This is the courtship of dippers, held during the short, dark days of winter. Heads stretched upward, next to one another in the icy stream, the birds twitter and exchange short bursts of song. They take off in synchrony, graceful dark shapes against the snow.

Filamentous algae form thick green strands undulating slowly in the current of a side channel. The September flood washed out many logjams upstream, releasing the wedge of twigs, pine needles, and organic ooze stored above each jam and creating a feast of nutrients for downstream aquatic plants and insects. The flood was a disaster for people living along the rivers at the base of the mountains, eroding banks that someone claimed as property and destroying houses and roads. For the rivers, the flood is a rejuvenating force, shifting channels across valley bottoms, recruiting tree trunks into the streams, leaving behind tangles of sediment and wood that create new habitat for riverine plants and animals.

In a valley bottom unconfined by bridge crossings, stabilized banks, or roads, animals can find marginal areas to wait out the force of the floodwaters. Secondary channels or zones of slower flow along the edges of pools or behind large boulders provide refuges from the tumult of water and sediment.

If the river channel and floodplain are sufficiently complex—messy—to provide these refuges during the flood, then insects and fish return to an apparently devastated river in abundance following the flood. The flood flushes finer sediment that has accumulated between cobbles and boulders, cleaning the streambed and improving the habitat for aquatic insects living on the bed, as well as the spawning grounds for fish.

The effects of the recent flood are not obvious in the beaver meadow. The broad, densely vegetated meadow dispersed the floodwaters barreling into it from the narrow channel upstream. All those willow stems bent under the swift waters but created enough frictional resistance to significantly slow the flow, allowing sediment being carried by the water to settle onto the meadow. Some of the floodwater soaked into the ground as the water moved more slowly through the meadow. Some of the water was stored in the isolated ponds that dot the valley bottom, remnants of former beaver dams and channels that have now shifted elsewhere. The meadow attenuated the flood, reducing the magnitude of the flood peak that reached the creek downstream.

Now, months after the drama of the flood, long, delicate ice crystals line the edge of the dark waters of the creek. I push gently on the crystals with a mittened hand and the ice shatters readily. It requires a leap of faith to understand that masses of ice have sculpted the ridges and peaks that rise so apparently massive and unyielding above me as I kneel beside the creek. But I have seen glaciers of dense blue ice dragging along house-sized boulders and collapsing in massive blocks that send small tsunamis across narrow bays. I believe in the power of ice.

The Setting

The waters of North St. Vrain Creek collect from tributaries that, on a map, resemble the fingers of a spread hand. Hunters Creek, Sandbeach Lake Creek, North St. Vrain Creek, Ouzel Creek, and Cony Creek all converge within a distance of less than two miles along the mainstem, and the beaver meadow lies just downstream, at the wrist. The hand rests palm upward, with the fingers extended into the air, for the creeks all descend steeply from the heights of rock and snow into the relatively flat main valley.

The creeks start as small, unnamed rivulets that form lines of aerated whitewater moving down the dark rock faces during summer. The little rivulets collect into lakes—Thunder Lake on North St. Vrain Creek, Bluebird and Ouzel Lakes on Ouzel Creek, Sandbeach Lake, and Finch and Pear Lakes on Cony Creek. Most of these lakes formed in depressions created by valley glaciers. Between the lakes are waterfalls where the creeks cannot cut into the rocks fast enough to create a gentle grade. Each of the creeks feeding the beaver meadow has a stepped profile, plunging steeply into the flat lakes, then flowing onward to drop abruptly over Copeland Falls,

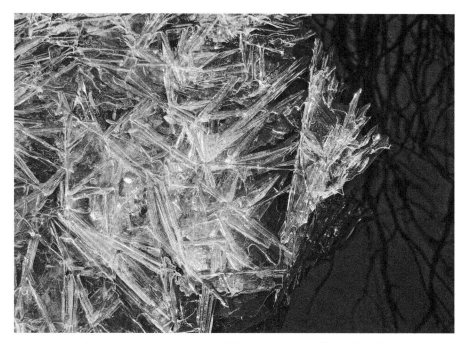

A close view of ice crystals along the edge of a beaver pond. Leafless willow branches are reflected in the still water at right.

Ouzel Falls, Calypso Cascades, and a dozen other, unnamed falls, before flattening again going into the beaver meadow. Each of the creeks descends from as high as 12,000 feet to about 8,300 feet in the meadow over a distance of some nine and a half miles.

The whole landscape is stepped. Go very far in any direction and you must go up or down. Water and sediment go down swiftly, carrying nutrients with them. Most of the length of each creek is a steep, narrow canyon with whitewater frothing rapidly along between big boulders or bedrock walls. At widely spaced intervals a slightly wider, gentler segment of valley interrupts these gorges. If the wider valley bottom is surrounded by old-growth forest, the big trees that fall into the creek can create logjams that span the channel. So many logs pile up in these relatively short, gentler portions of the valley that the backwater behind each logjam forces flow over the creek banks. Smaller, secondary channels branch and rejoin across the valley. These portions of the creek create a temporary resting place for water, sediment, and nutrients. They are the biological hot spots along the creek, creating habitat for a greater variety of plants and animals.

If the wider valley bottom is sufficiently wet to support the deciduous trees that can tolerate damp or saturated soils—willow, aspen, river birch, cottonwood— these trees provide beaver food and the site becomes a beaver meadow. As long as

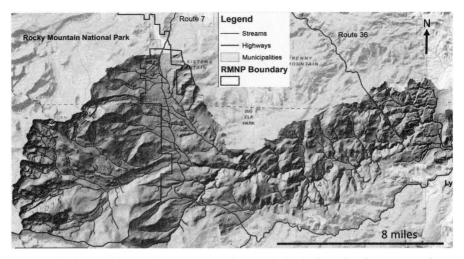

Shaded relief map of the North St. Vrain Creek watershed. The boundary between Rocky Mountain National Park and Roosevelt National Forest bisects the basin, and the town of Lyons is at the base of the watershed, along the mountain front. The western boundary of the watershed is the Continental Divide. A color version of this figure is included in the insert section.

they actively maintain dams, the beavers also actively maintain this wet meadow, providing a home for themselves and for a rich array of other plants and animals.

Sometimes, however, the beavers cannot make a go of it. Perhaps there is not enough space on the valley bottom to support the large stands of deciduous trees necessary to feed a beaver. Perhaps the valley is so narrow that it concentrates the snowmelt floods into a powerful surge that rips out the beaver dams. Whatever the reason, abandoned, breached beaver dams partly span many of the steeper portions of the creeks. Daredevil beavers built some of these dams immediately above tall waterfalls. The most persistent beaver meadows occupy the wider valley segments lower in the drainage basin, where events through geological history have shaped a valley ideally suited for beavers.

The Rocks

The western boundary of the North St. Vrain Creek watershed divides water flowing to the Pacific Ocean from water flowing to the Atlantic. This is the spine of North America, a landscape resulting from geological violence where two tectonic plates crashed together and thrust upward into mountains along the convergence zone. Portions of the western plate accreted onto the eastern plate and now the plate boundary is far to the west, where the violence continues in abrupt jerks along the San Andreas Fault and explosive eruptions from the Cascade Volcanoes.

Moving the boundary to its current location took a long time. Multiple times the Rockies were uplifted and then worn down by weathering of bedrock into sediment that wind, water, and ice carried to lower elevations. The direction and rate of movement of the great tectonic plates that meet at the western boundary of North America have changed repeatedly during the past 300 million years. Each change triggers a new episode of mountain building or a quiet period during which weathering and erosion once more dismantle the mountains. The most recent episode of renewed uplift and mountain building ended about 40 million years ago.

Raising the mountains effectively shakes up the rivers, steepening their courses and setting them to cutting downward as fast as they can. Water and the sediment it carries are the tools of river erosion, opposed by the strength of the bedrock forming the mountains. The appearance of the mountains and river valleys at any point in time reflects the balance between these driving forces of erosion and the resistance of the rock. Precipitation feeds the rivers and allows them to cut downward, but the Rockies are now the interior mountains, far from any ocean that can serve as a source of the water vapor that feeds precipitation.

Water vapor coming inland from the Pacific Ocean rides an enormous atmospheric roller coaster as it crosses the Coast Ranges and every other line of mountains between the coast and the eastern face of the Rockies where North St. Vrain Creek lies. With each rise, the air cools and some of the vapor condenses and falls as rain or snow. By the time the air reaches the Southern Rockies, there is typically not much water vapor left. The best chance for a big snowstorm is when Arctic air funneling southward along the eastern edge of the Rockies meets moist air moving east from the Pacific or northwest from the Gulf of Mexico. In spring and summer, air coming inland from the Gulf carries abundant moisture, but much of this is wrung from the air at lower elevations in the foothills. The big rains and snows are few and far between at the elevation of the beaver meadow. This is primarily a dry landscape by world standards. Each year, only 24 to 47 inches of precipitation fall at the highest elevations, and the abundance of water defines survival for plants and animals.

The Rockies change slowly. Their location in the continental interior limits inputs of tectonic energy from Earth's interior, as well as energy for erosion in the form of atmospheric moisture. The pace of landscape evolution has been set most recently by the advances and retreats of the valley glaciers during the past two million years.

A satellite image clearly reveals where valley glaciers scooped steep-walled troughs from the resistant granite bedrock surrounding North St. Vrain Creek. The highest peaks were never covered by glaciers, but ice has worried their flanks for as long as the peaks have existed. Isolated patches of ice chip away at the tough crystalline rock, infiltrating as water during each thaw, then freezing and expanding to shatter the rock. Like wrinkled skin that reflects a long life of exposure to sun and wind, the hairline cracks of joints in bedrock reflect the long history of uplift and erosion. Joints form as pressure is released during uplift or removal of overlying rock. Erosion then enlarges the joints. Joints are the lines of weakness by which water, ice, plants, and gravity can break the rock apart.

The highest peaks form in massive bedrock where the joints are farthest apart. This resistant rock rises to summits 12,000 to 14,000 feet above sea level. Surrounding the North St. Vrain drainage with a rocky rampart stand, clockwise, Mount Meeker, Longs Peak, Chiefs Head, Mount Alice, Isolation Peak, Ogalalla Peak, and St. Vrain Mountain. Across the divide to the north rise rock formations named Keyboard of the Winds, The Notch, The Loft, The Cleaver, and Storm Peak.

The Ice

Something there is, though, that doesn't love a mountain. Or, rather, lots of things—glaciers, rivers, and gravity combine to dismantle the mountains. The amount of solar radiation reaching Earth's Northern Hemisphere reflects orbital characteristics of the planet that vary over different lengths of time. Two million years ago, these variations combined to reduce the solar energy reaching the great landmasses of the Northern Hemisphere. Ice sheets of continental size spread across Eurasia and North America. The Southern Rockies of Colorado were not covered by the great Laurentide Ice Sheet that spread across Canada and down into the northern fringes of the continental United States. Colorado was too far south and too far from moisture carried in the atmosphere from the Pacific Ocean. Only isolated valley glaciers formed in the Colorado Rockies. These smaller tongues of ice flowed downward from collection points near the Continental Divide, following existing river valleys.

Although small compared to the continental ice sheets, the valley glaciers were large and powerful enough to leave an impressive signature on the landscape. The beaver meadow on North St. Vrain Creek lies at just over 8,300 feet in elevation. The valley walls on either side rise steeply for the first 500 feet, then rise much more gently to the bases of the steep peaks. The first 500 feet are the sides of the lateral moraines left by the glaciers.

Glacial ice itself is too soft to erode anything, although the ice can break apart bedrock by melting into cracks and then refreezing. Ice melts and deforms readily under pressure, so ice squeezed beneath the weight of a glacier simply melts and flows over and around obstacles in its path. Meltwater that freezes back into the glacier, however, can include chunks of rock that are then dragged along with the flowing ice. It is these abrasives frozen into the ice that accomplish glacial erosion, transforming the glacier into a giant sheet of sandpaper moving across the terrain.

A glacier is more like a flowing slushy than a block of ice. The glacier flows primarily on a layer of meltwater between the ice and the underlying rock, and meltwater

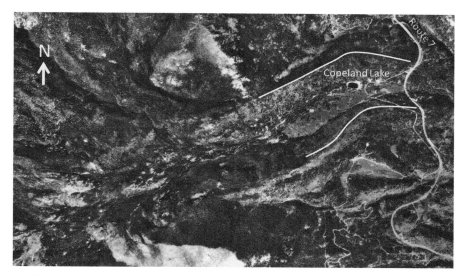

Satellite image of the area around the beaver meadow in 1999. Route 7 is at right. The beaver meadow is the long, flat area just south of Copeland Lake. The lateral moraines (solid white lines) parallel the North St. Vrain valley to north and south. Flow direction is to the right. Image courtesy of Google Earth.

is abundant within the ice, forming sediment-laden streams that twist sinuously into and through the glacier. Sediment within the ice is continually being exchanged with the adjacent terrain. Rock falls from the valley walls above the glacier, and wind deposits silt and clay on the surface of the ice. Sediment within the ice or moving with the meltwater streams is deposited in front of, beneath, or beside the ice.

The deposition of sediment accelerates as a glacier loses mass. The front of a retreating glacier can be difficult to identify because the ice is so thickly buried by sediment that the glacier front looks like a pile of rock. When a glacier retreats, it does so in fits and starts. The ice stagnates at one position for a time, retreats rapidly, then stagnates again. "Stagnate" is misleading, however, as is "retreat." Even a retreating glacier continues to flow downslope. A glacier in retreat does not have enough mass to flow as far downslope, so the farthest extent of the glacier decreases through time. A stagnant glacier continues to flow and carry sediment downslope, leaving behind a large accumulation of sediment when it does retreat. This sediment can form a mound across the former front of the glacier known as an end moraine. Because glaciers retreat in fits and starts, a single alpine glacier can leave behind a long trail of end moraines like footprints. The end moraine farthest down-valley is the terminal moraine. Sediment released by the melting ice can also create lateral moraines in the form of a mound along each side of the valley parallel to the flow direction. The top of the lateral moraine indicates the maximum height of the glacial ice.

View of the beaver meadow from the Sandbeach Lake trail. Lateral moraine at rear (down valley to the left). A color version of this figure is included in the insert section.

Glacial moraines cradle the North St. Vrain beaver meadow. The rocky slopes of the lateral moraines drain water readily into the valley bottom, and conifers that prefer dry soils grow on these slopes. The terminal moraine stands just downstream. Although North St. Vrain Creek has cut a narrow breach into this mound of rock and soil, the moraine creates a sill along the valley bottom. The sill trapped finer sediment released as the glacier melted and retreated farther up-valley, and now the sill limits the ability of the creek to lower the entire valley bottom. Because of the sill, the creek just upstream meanders across a broad, gentle, wet valley bottom that is perfect for a beaver to call home.

After the Glaciers

Imagine this valley 15,000 years ago. The ice front has retreated up the valley from the terminal moraine, but not very far up. The mass of ice creates a strong temperature contrast with the adjacent bare rock. Forceful winds frequently sweep down the valley. Melting ice has dumped large quantities of sediment across the valley. Surging winds pick up silt and clay and transport them long distances. Meltwaters rework sediment into alluvial fans and braided rivers. Individual channels shift constantly across these fans as sediment deposited in one place builds up the local elevation and the flowing water abruptly switches to a lower portion of the fan. Pulses of water and sediment coming off the melting ice keep the landscape too mobile for plants to gain much of a roothold.

Eventually the glacier front retreats far enough up valley that the pulses of water and sediment are attenuated by distance. Pioneering plants start to colonize the valley bottom. Tundra comes in first—grasses, sedges, lichens, mosses, clubmosses, and dwarf flowering perennials. With time, seeds of spruce and fir germinate and grow into a subalpine forest, and then give way to a montane forest of predominantly pine. By about 7,000 years ago, warm temperatures allow the timberline to extend to higher elevations than it reaches today, but subsequent cooling again limits the elevations reached by trees.

Animals follow their food sources as plant communities cover the newly exposed terrain. Some of these are animals almost unimaginable today. Construction in 2011 exposed the buried shoreline of a small glacial lake in Snowmass, near the town of Aspen, about 80 miles southwest of the North St. Vrain beaver meadow. Between 150,000 and 130,000 years ago, the environment surrounding the Snowmass glacial lake alternated between tundra and forest. An array of giants left their bones beside the lake—mammoths, mastodons, giant bison, camels, Pleistocene horses, ground sloths, and giant beavers—the fauna of the Pleistocene Southern Rockies.

The Pleistocene valley glaciers of the Southern Rockies advanced and retreated at least three times during the period between two million years and 10,000 years ago. Each period of retreat allowed plants and animals to reclaim terrain lost to the ice. Presumably, the creatures of the earlier interglacial period at the Snowmass site also roamed the site of the North St. Vrain beaver meadow as the latest Pleistocene glaciers retreated, then gave way to smaller species such as today's elk, bighorn sheep, and mule deer.

At some point, modern beavers moved into the valley of North St. Vrain Creek. Exactly when this happened is not known. Beaver-pond sediments elsewhere in the national park date to at least 5,000 years ago and likely older. That means the beavers survived the Neoglacial.

No masses of ice comparable in size to the Pleistocene glaciers have formed at temperate latitudes since the start of the Holocene period of geological time, 10,000 years ago. Smaller alpine glaciers formed and melted, however, as Holocene climate alternated between cooler and warmer periods. Glaciers advanced during the Neoglacial period about 4,000 to 3,000 years ago in the Colorado Front Range and during the Little Ice Age of circa 1550 to 1850 AD. Glaciers retreated during the Altithermal period of 7,000 to 4,000 years ago and the Medieval Optimum of circa 950 to 1250 AD. Glacial retreat turned into a gallop during the latter half of the 20th century, and no glaciers now remain in the North St. Vrain Creek drainage.

Presumably, beavers in the Southern Rockies retreated to lower elevations during cool intervals and then recolonized upper elevations during warm intervals.

Studies of beaver colonies in Yellowstone National Park indicate that the beavers abandoned smaller channels during warm, dry periods when the smaller channels probably stopped flowing. Beavers may have been able to live in the North St. Vrain beaver meadow throughout the Holocene. Glaciers never again advanced down to this elevation during the Holocene, and the creek is large enough to have remained flowing even during warm intervals. Certainly by about 5,000 years ago, the types of vegetation communities present in the region today were in existence and the beavers were there with them.

Whenever they arrived, the beavers set to work. Within the valley template created by tectonic uplift and glacial ice, the beavers built dams and rearranged the movement of water, sediment, and nutrients to serve their own purposes, in the process creating a distinctive beaver meadow that provided habitat for dozens of other plants and animals. The beavers had engineered their own ecosystem by the time people arrived, and the animals came to represent the essence of that ecosystem for some people. As Grey Owl (Archibald Belaney) wrote in his 1935 book *Pilgrims of the Wild*:

> I had travelled nearly two thousand miles by canoe through a reputed beaver country [Ontario] to find only here and there a thinly populated colony, or odd survivors living alone. . . . The beaver were going fast; in large areas they were already gone. Was this then, to be the end? Beaver stood for something vital, something essential in this wilderness, were a component part of it; they *were* the wilderness. With them gone it would be empty; without them it would be not a wilderness but a waste. (pp. 47–48)

For other people, more prosaic, the beavers were simply another resource to be used to depletion or a competitor to be kept at bay. The resource view dominated until the beavers were nearly extinct at the end of the 19th century, and the competitor view continues to be too common today.

February

About Beavers

The Meadow in February

February: when I first notice the days begin to stretch out after the tight curl of December and January. February: the month for creating new beavers. Somewhere sheltered from the cold light of sun reflecting off snow, in a bank den or a lodge, perhaps even in the water, the beavers are mating. The North St. Vrain beaver meadow is good habitat, and the adult pair will create another litter of two to four kits, allowing the current kits to become yearlings, and pushing the current yearlings out of the colony to find new homes and their own mates.

I enter the meadow from the northern side on a mild day when a steady breeze seems to keep pace with the scattered clouds moving overhead. Just as I start to descend the slope into the meadow, I flush a moose resting beneath a big spruce. I am close enough to see the coarse, thick fur along the moose's spine bristle with alarm and to note the scars where the animal dropped its antlers after the autumn rut. The alarm is only momentary. The moose ambles down the slope and into the meadow, steadily browsing willow stems as it moves. The stems have a hue of burnt yellow, and each bud is clearly visible, although not yet swollen.

The past few days have been mild, and patches of bare ground show along the south-facing slopes and under the big conifers that border the beaver meadow. In the meadow itself the snow remains sufficiently deep to keep navigation easier by filling the pitfalls of the meadow—the winding canals, one beaver wide, and the steep-sided holes that the beavers dig to create air exchange near their dens. The snow bears abundant witness to the activity of the meadow. The widely spaced leaps of snowshoe hare tracks create diagonals between the creek and the conifer forests adjacent to the meadow. The single, precise line left by a fox has melted in the warmth of midday and refrozen into ice casts. The delicate, leaping tracks of a mouse barely indent the snow where the surface is powdery. The moose has left deep, snow-covered divots. Neat brown piles of moose poop linger at the surface long after the accompanying tracks have vanished.

I snowshoe beside the channel, stooping and struggling through patches of dense willows, wondering at the mysterious ways of the creek in winter. A side branch of the creek disappears under a snow bank for several meters, then returns abruptly to the surface. Ice forms filigree patterns along the water's edge, translucent between the white, opaque snow and the transparent water. The flowing water massages the ice, leaving it bulbed, layered, and contoured. Streaks of pale sand lie among the brown cobbles in the streambed, traces of the September flood. The cobbles are coated with green or brownish-green algae. Turning over even a few cobbles reveals the pebble-and-twig cases of caddisflies attached to the undersides of the rocks, or discomfited mayfly and stonefly larvae wriggling away. Decaying pine needles, willow leaves, and twigs form dark windrows along the margins of the flow where the current is slower, food for the aquatic insects that collect bits of organic detritus. As I continue downstream, the thick, continuous layer of organic detritus on the streambed indicates the approach to an intact beaver dam as much as the nearly still water.

The beaver dam is small, about 3 feet tall and 16 feet wide. This is not the engineering marvel that beavers are capable of—a beaver Great Wall that spans distances of thousands of feet and holds thousands of gallons of water perched several feet above the main valley floor. But in its way, the little dam is still marvelous.

Worth a Dam

The sound of running water can trigger the urge to build a dam in listening beavers. Fortunately, it's not as simple as that, or the so-called bank beavers that burrow into the banks of rivers too large to dam would presumably spend their lives frustrated and neurotic. In her marvelous book *Lily Pond*, an account of watching a colony of beavers in New York over four years, Hope Ryden describes how beavers build dams to obtain the result of controlling the water level in their environment. Ryden writes of beavers assiduously repairing and checking dams despite the lack of apparent leaks or overtopping, as well as beavers blithely ignoring water running over or leaking through their dams when water level in the pond is of sufficient height.

When the water level is not satisfactory, beavers go to work, and patient scientists have watched the process of a dam taking shape. First come the foundational materials: rocks, clay, or pieces of wood pushed into a ridge perpendicular to the flow. Sometimes the beaver takes advantage of an existing partial barrier, building the dam on top of the boulder lobe left along a steep channel by a debris flow. Once the foundation is laid, the beaver adds structure, anchoring leafy branches, peeled branches, and just about anything else likely to help to make the dam an impermeable barrier. I have seen dams with clay plastered over the core of interwoven, thick branches; dams with cobbles piled up against them; and a dam that incorporated a full rack of moose antlers. Most beaver dams are about 4 to 6 feet tall, although at least one overachieving beaver colony built a dam more than 15 feet tall in Wyoming. Enos Mills described a dam that he measured as being 2,140 feet long on the Jefferson River near Three Forks, Montana. In 2007, Canadian ecologist Jean Thie noticed a beaver dam 2,800 feet long on a satellite image of Canada's Wood Buffalo National Park.

One beaver lodge and pond near Long's Peak were occupied continuously for 70 years, but mostly the beavers move on after a few years, building a new dam elsewhere within the beaver meadow and starting again the succession from flowing stream to pond to marsh to meadow.

A beaver, which normally moves about in an awkward waddle when out of the water, can walk upright on its hind legs while carrying dam-building materials held against its chest with its chin and front legs. The beaver's short, heavily clawed front feet facilitate digging yet are sufficiently dexterous to fold individual leaves into its mouth or to rotate slender stems like a cob of corn as the beaver gnaws off the bark. A beaver uses the large, webbed paws of its hind feet to kick away the dirt it digs out.

Beavers do not have to build dams, but water ponded behind the dam serves as both protective moat and refrigerator for the beaver colony. Moat, in that beavers waddling on land transform abruptly into graceful, efficient swimmers in the water and use underwater entrances into lodges that contain a nest area above the water level. Refrigerator, in that beavers create winter food caches by submerging piles of branches near the lodge.

Beavers dig canals across the valley bottom, as well, to make it easier to move about and bring food back to the lodge. These canals are mostly less than 75 feet long, although some have been measured at 450 feet or longer. Some canals are dug through otherwise dry land on the floodplain. Other canals are dug into the bed of shallow ponds to ensure sufficient swimming depth during dry periods, similar to the human practice of dredging rivers for navigation.

If beaver dams served only the beavers, they would be an interesting facet of the animal's behavior. What is marvelous about beavers and their dams is the resulting transformation of the landscape. Like nobility of the animal kingdom, beavers can lay claim to two titles. As a keystone species, beavers exert an outsize influence on the physical appearance of the landscape and on the ability of other species to inhabit the landscape. As ecosystem engineers, beavers control the availability of resources for other species. Simply put, beavers build dams, which store water, creating ponds and wetlands. Everything else flows from this.

Upstream view of a beaver dam on a side channel in the North St. Vrain beaver meadow. A color version of this figure is included in the insert section.

About Beavers

February in the North St. Vrain beaver meadow is the lean season for the beavers. Beavers eat a variety of plants, from the twigs and bark of woody plants such as willow, aspen, alder, birch, and cottonwood, to grasses, sedges, and herbaceous plants such as rooted aquatic species. Their summer diet features salad—up to half of what beavers eat is forbs and grasses, but during the winter they mostly subsist on the cache of shrub and tree branches that they stashed underwater somewhere near their lodge during the autumn. An adult beaver needs anywhere from more than a pound to four pounds of twigs and bark each day, but even this is not enough. Beavers store fat in their tails and metabolize this surplus during the winter. They also use less energy and do not grow during winter, when most of their activity occurs closer to the den than during other seasons.

Undoubtedly one of the reasons that beavers can occupy areas from northern Alaska to northern Mexico is their ability to adapt to diverse foods. In *Lily Pond*, Hope Ryden describes a beaver colony that survived through the winter primarily on the roots of water lilies that they dug up from the pond bottom. After more than one winter of intensive lily-root harvest, the lily population was so reduced that the beavers had to move to a new pond, but they did make it

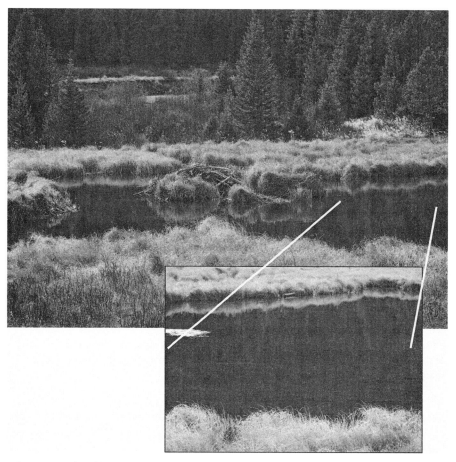

A beaver pond and lodge in Colorado, with inset view showing the beaver canal dug into the bed of the pond as a darker band across the center of the pond. A color version of this figure is included in the insert section.

through more than one cold winter without much of a cache of woody branches and trunks.

A beaver can exceed 40 inches from the tip of its nose to the tip of its tail. Adults typically weigh more than 50 pounds. Although a Methuselah-beaver of 21 has been found, most beavers live about 10 to 12 years, and a good many live less than 5 years. Up to half of beaver kits may die during their first six months. Adolescence and young adulthood are also a particularly dangerous time. Young beavers typically leave their natal colony at two years of age to seek new territory, and this is when they are most vulnerable to competition with established beavers, predation, and accidents. During one study of beavers in Massachusetts, where the animals have been increasing because of restrictions on fur trapping, 78% of the documented

Part of a beaver's winter food cache. In this view, with snow along the bank at the bottom of the photograph, the beaver has recently broken the ice to access the gnawed, peeled branches that it cached under water the preceding autumn.

mortalities were among juveniles dispersing from their natal colony. These juveniles mostly moved during April to June and traveled an average of 2.7 miles along rivers before settling down again.

If habitat is available near its natal colony, a beaver will settle there. Many studies of dispersal indicate that beavers prefer to remain within about three miles of their home. Like some younger human adults, beavers also sometimes fail to launch. Hope Ryden described beavers returning to their natal colony after a year or more away, joining the family group for a time, and then moving on again.

Dispersing young beavers sometimes move over longer distances in stages spread across several years if food is limited near existing colonies. In areas with high densities of beavers, the youngsters are more likely to attempt to colonize habitat of lower quality, putting them at greater risk of death. If a beaver can survive its first five years, it is more likely to enjoy a full beaver lifespan.

Young beavers moving on into the wide world must find suitable habitat, but they must also contend with established beaver colonies. Beavers are highly territorial. Both sexes scent-mark the boundaries of a colony's territory, typically about a quarter-mile above and below the primary pond, which limits colony density to one or two colonies per mile of stream even in prime habitat. Prime habitat for a beaver is a wide valley sloping gently downstream. Beavers make scent mounds by excreting scent from their castor and anal glands onto piles of mud and sticks that they carry and place about their territory to create a scent fence. A single anal secretion from a beaver can include as many as a hundred different chemical compounds, and each beaver has a distinctive scent, analogous to the famous uniqueness of human fingerprints. Beavers do not see well, so they use scent and excellent hearing to assess the world around them, and scent in particular to distinguish relatives from intruders. Scientists and resource managers use attractive scents to manipulate the establishment of new beaver colonies, concocting gloppy paste that acts as a siren song for beavers, luring the animals into live traps and suitable new habitat.

The existence of scent fences is one of those facts uncovered in scientific studies that remind me of how fundamentally ignorant and oblivious I am when exploring a beaver meadow. Focusing as intently as I can on the surroundings, I will never smell a scent mound or be stopped in my tracks by a scent fence, just as I will never hear a vole foraging beneath the snow in the way that a hunting fox or coyote can. I wonder how different this landscape appears to a beaver, who smells a scent fence and decides to turn back, or hears the sound of running water and feels the urge to build a dam that will create a pond and support a colony.

The average colony includes 5 or 6 beavers, although a colony can contain 12 individuals. Beavers are monogamous for life, and only the adult pair in a colony reproduces, although females generally do not bear a litter every year. The remainder of the colony consists of the offspring—the two to four young of the year, known as kits, and the young of the previous year, known as yearlings. If beavers were human, we would describe their society as matriarchal. The adult female emerges from the lodge first most of the time, and she leads the colony in building and maintaining the lodge and the dams, as well as in building the winter food cache.

A colony requires a minimum area of almost 10 acres of willows to persist indefinitely. This size of larder provides the beavers with enough food but also allows the willows to regenerate between beaver harvests. Willows and some other species of woody riparian plants have their own evolutionary adaptations to the slings and arrows of life in a beaver meadow. Even when severed near the ground by beaver gnawing or by logs borne on floodwaters, a willow stem can regenerate from the roots. Broken bits of plant carried downstream can take root where they come to rest. Willows can also send out rhizomes—below-ground stems that surface some distance away from the parent plant, allowing the willow to regenerate asexually. And willows release seeds that wind transports to new germination sites. These multiple strategies allow the willows to persist despite the presence of voracious beavers.

A single adult beaver can cut 200 to 300 trees a year, most of which will be chosen within 100 feet of the water's edge. At some level, a beaver has to cut this many trees because its incisor teeth grow continuously. The upper incisors form a broad semicircle within the skull, starting far back toward the brain and arcing up and out to create the prominent, orange-hued chisel for which beavers are famous. The chiseled edge remains sharp by grinding the upper and lower incisors against each other, and the orange enameled front surface provides the cutting edge to fell trees and peel bark.

Beaver skeletons are massive compared to those of other mammals of similar length. The thick, heavy skull and jawbone provide a strong foundation for the animal's large incisors and are able to withstand the physical stresses of jaw muscle contractions as the beaver cuts aptly named hardwood tree trunks.

Viewed as the product of evolutionary adaptation, a beaver is a marvelous tool for living in a stream and building dams. The animal can remain underwater for up to 20 minutes, swimming powerfully using its tail as a rudder and propeller and its webbed hind feet for kicking. Swimming under ice during the winter presents no problems. The beaver's ears and nostrils close underwater. A beaver even has fur-lined inner lips that close behind its teeth so that it can chew underwater in comfort. A special membrane over the beaver's eyes acts like swim goggles, allowing the beaver to see underwater. The animal's eyes are even better than swim goggles,

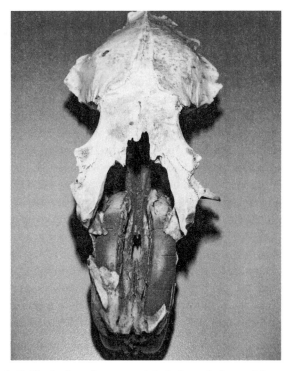

Top of a beaver skull. The incisors form a semicircle from the base of the eye sockets to the protruding tips at the front of the animal's mouth. A color version of this figure is included in the insert section.

though, because specialized eye muscles correct for the refraction of light underwater so that underwater objects do not appear distorted. The beaver can keep its fur water repellent by rubbing oil from two sacs on each side of its urogenital openings into the fur. Of course, it's hard to rub your own back. A beaver grooms itself wherever it can reach but relies on other family members to groom its back fur. The grooming also maintains a layer of air in the fur that helps to conserve body warmth. The tail is filled with minute blood vessels and specialized circulatory features that help the beaver conserve heat during winter and radiate body heat during summer.

Beavers also conserve heat during winter by minimizing their activity and spending more time in the lodge or bank den. A bank den is typically dug under a large tree or shrub, the roots of which help to support the roof of the den. The den includes a nest above water level, an underwater entrance, and small holes to the surface to keep the interior air fresh. A lodge can be built on land or within a pond or shallow lake. The structure is composed of debarked sticks and mud, with air holes and underwater entrances similar to those in a bank den. Even in summer, beavers spend much of the day within the bank den or lodge, emerging to forage and engage in other activities at dawn and dusk (an activity pattern known as crepuscular), as well as during the night.

This schedule of activity partly reflects the need to avoid predators. Wolves, coyotes, and mountain lions have historically been the major predators—other than humans—but beavers can also be killed by bears, wolverines, river otters, lynxes, bobcats, and minks. A study undertaken during 2012–2013 in Flanders, the northern region of Belgium, suggests that beavers maintain their nocturnal schedule even in the absence of predators. Beavers living within a nature reserve in Flanders would seem to enjoy the good life. Humans do not hunt in the preserve, no natural predators of beavers remain in the area, and even the domestic dogs are walked on a leash and not allowed to roam freely, yet still the beavers remain circumspect. The scientists conducting the study, Kristijn Swinnen and her colleagues, used camera traps to record the activity of the beavers and found that the animals remained crepuscular and nocturnal even without the need to avoid predators. They inferred that the beavers had not yet relaxed their vigilance but instead continued to be influenced by the "ghosts of predators past." This strikes me as a 21st-century version of what a 19th-century writer might have described as an ancestral memory and a 20th-century writer as an instinct. Whatever we call it, the beavers remain cautious.

A beaver can slap its tail loudly on the water surface to give a warning against predators and other dangers. Beavers typically express frustration toward other colony members through pushing matches akin to Japanese sumo wrestling. As Hope Ryden notes, use of their powerful front teeth to inflict wounds on other family members would be extremely dangerous among beavers, particularly when they spend long winters nestled close to one another in a lodge. Teeth are used as weapons only against outsiders and then only as a last resort.

Ecologists have described seven vocal sounds for beavers, but adults typically use only three outside of the lodge—whining, hissing, and growling. In his 1914 book *The Romance of the Beaver*, Arthur Radclyffe Dugmore wrote a particularly evocative description of "beaver language, which sounds like a strange subdued mixture of children's voices and very young pigs squealing, varied now and then by a puppy's cry" (p. 169). Kits tend to be more vocal, like juveniles of most species. Kits beg for food, whine when food is at risk of being taken away, and vocalize to initiate grooming and play. Describing the vocalizations of kits, Grey Owl wrote in his book *Pilgrims of the Wild*: "Their voices were really the most remarkable thing about them, much resembling the cries of a human infant, without the volume but with a greater variety of expression" (p. 33). Who wouldn't be attracted to an infant with more variety and lesser volume of sound?

Hope Ryden also writes of yearlings and adults vocalizing, particularly when another colony member attempts to take away a particularly choice morsel of food. In some of the most endearing passages in *Lily Pond*, she describes reassuring herself of the beavers' survival during the winter by listening for the sounds of contented beaver murmuring from within the ice-bound lodge.

Adult beavers play, as well. Enos Mills described "a general frolic" that lasted more than an hour one autumn morning: "They raced, dived, crowded in general mix-ups, whacked the water with their tails, wrestled, and dived again . . . the merrymakers splashed water all over the main pond before they calmed down and in silence returned to work" (pp. 156–157). How can anyone help liking engineers that engage in general frolic?

The Family Tree of Beavers

Part of the reason that willows and other riparian plants in North America and Eurasia can withstand being harvested by beavers is that the plants have had ample time in which to adapt to beavers: beavers have a long history on these continents. Beavers are the largest rodents in North America. Rodents first appear in the fossil record between 45 million and 38 million years ago, when non-avian dinosaurs disappeared into the evolutionary darkness known as extinction.

More specifically, beavers are castorids, a family of animals that now includes only the two species of beavers. The family has an impressive history, however, with at least 30 different genera in the fossil record. North American beavers are members of the species *Castor canadensis* and Eurasian beavers are *Castor fiber*. Both species have existed for at least 1.9 million years.

Castorids appear in the fossil record in North America (genus *Agnotocastor*) and Asia (genus *Steneofiber*) during the late Eocene and Oligocene epochs of geologic time—approximately 50 million to 30 million years ago. Castorids seem to have evolved in North America first and then spread to Eurasia. These proto-beavers had teeth that were not well suited to gnawing wood, a behavior that evolved later, but they do appear to have been adapted to semi-aquatic living. By about 20 million years ago, animals of the genus *Palaeocastor* appear in Nebraska. These beaver ancestors were about the size of a modern muskrat and dug burrows up to eight feet deep into riverbanks and the shorelines of ponds and lakes.

By five million years ago, beavers were cutting wood and building homes. Beaver Pond, a site on Ellesmere Island at 78°N in Canada's High Arctic, is rich in cut sticks and fossil remains of beavers four to five million years old. These beavers were only about a third of the size of contemporary beavers. Some of the ancient species of beavers in North America weighed less than three pounds; others were nearly 10 feet long.

Many types of terrestrial mammals reached their zenith, in terms of size, during the Pleistocene epoch when the great continental ice sheets were advancing and retreating in the Northern Hemisphere. North America was home to saber-toothed cats, giant ground sloths, mammoths, mastodons, camels—all the animals found at the Snowmass site—and giant beavers. Eurasian Pleistocene beavers belonged to the genus *Trogontherium* and the North American giants were *Castoroides*. The giant *Castoroides ohioensis* left fossil remains as far south as Florida and as far north as Old Crow in the Yukon Territory of Canada.

Castoroides may not have been the Einstein of the animal world. These beavers had a body the size of a modern black bear, reaching a body weight of about 500 pounds, but a brain only slightly larger than the modest size of modern beavers. They did have the other marks of a modern beaver: a body shape adapted to swimming and those efficient chisel teeth. The giants coexisted with the modern *Castor canadensis* for a time but then went extinct about 10,000 years ago.

Castor canadensis appears to have followed the retreating continental ice sheets northward, colonizing newly exposed lowlands in southeastern New England by 12,000 years ago and in British Columbia by 10,000 years ago. Modern beavers have been present for even longer at sites in Oklahoma, New Mexico, and other areas south of the ice sheets.

Although appearance and behavior might suggest that beavers are closely related to muskrats and South American nutrias, these animals are actually distant cousins. The closest living relative of modern beavers is the squirrel, which has leg bones similar in shape to those of beavers.

Subspecies of North American beavers likely existed prior to intensive hunting by humans. Beavers inhabited islands along the Atlantic and Pacific coasts, ponds and marshes of the boreal regions of Alaska and Canada, swamps of the Deep South, and even perennial rivers of the desert Southwest: the geographic spread of *Castor canadensis* suggests that subspecies were likely. Once the animals were hunted nearly to extinction, however, and then reintroduced to some areas by humans, as well as reintroducing themselves by individual migrations, these regional distinctions were lost. Today, beavers look pretty much the same everywhere in North America, although they tend to be slightly smaller and lighter in color at the southern end of their range.

The Southern Colonists

People have also introduced beavers to regions where the animals were never native. In 1946, beavers were brought to the very tip of South America, to the Argentinean portion of the island of Tierra del Fuego near Fagnano Lake. The Argentine government had decided to "economically enhance" the remote, sparsely populated region by introducing exotic furbearing animals—beavers, muskrats, and minks—that could then be trapped for profit.

At a conservative estimate, there must be at least 100 examples worldwide of why introducing an exotic species is a very, very bad idea. Governments in North and South America, Australia, and New Zealand, in particular, now spend billions of dollars trying to control the spread of plants and animals deliberately or accidentally introduced from other continents, typically Eurasia. Some introduced species do not cause problems, but a great many become what are known as invasive exotics. Freed from the constraints inherent in the region where the species evolved—typically competition, predation in the case of animals, or grazing or insect infestation in the case of plants—these species have a proverbial field day, spreading aggressively into their own new worlds. In some cases, the invasive exotics outcompete native species, driving the natives toward extinction. A more insidious effect occurs when the invasive exotics change the very environment in a manner that largely eliminates the natives, as in the case of cheat grass.

Invasive exotic cheat grass reached North America in grain shipments from Eurasia, spread across the continent along railroad lines, and then filled in the gaps when its seeds lodged in animal fur and human clothing. Cheat grass burns very well. The grass spread as an understory through native sagebrush communities in the interior western United States, facilitating repeated range fires. These fires kill the native sagebrush, in turn destroying the habitat for sage grouse, which require the plant structure and density provided by sagebrush. Cheat grass now occupies thousands of square miles of western North America.

Beavers in South America are having an effect similar to that of cheat grass. The Argentine government initially released 25 mating pairs of beavers. The beavers explored the water-rich environment of Tierra del Fuego and found it good. By the 1950s, the rapidly expanding population had entered the Chilean portion of the large island. By the 1960s, enterprising immigrants had crossed ocean channels to the adjacent Chilean archipelago, where they colonized several other large islands. The Southern Hemisphere beaver colonies increased to greater spatial densities than are common in North America, and they now inhabit nearly every available watershed on most of the islands. Ecologists fear that the beavers will cross to the mainland and work their way northward. The region is so remote and sparsely populated by humans that this may already have occurred.

The problem with this example of beaver enterprise is that the native ecosystems of Tierra del Fuego have not developed with beavers. The archipelago at the southern tip of South America has a relatively low diversity of native plants and animals. Primary plant communities include *Sphagnum* peatlands, cushion bogs, rush wetlands, and *Nothofagus* forests. *Nothofagus*, sometimes described as a beech, is a broad-leafed tree characteristic of the Southern Hemisphere. *Nothofagus* is to southern South America what *Eucalyptus* is to Australia: the plant genus that dominates forests across the continent. Tierra del Fuego contains only three major tree species, all of which are *Nothofagus* and all of which are eaten by the introduced beavers. Not having evolved with beavers, *Nothofagus* trees lack the defensive mechanisms and reproductive strategies that North American trees have evolved to survive the effects of beaver foraging and flooding. *Nothofagus* forests regenerate from sapling banks—which the introduced beavers eat—rather than from the seed banks so important to North American riparian trees, because seeds do not persist for long in the wet soils of Tierra del Fuego. *Nothofagus* forests also appear to have trouble regenerating when they are subject to systematic or permanent stresses of the type created by beaver foraging and flooding. Only one species of *Nothofagus* is adapted to the boggy conditions associated with beaver meadows. As a result, beavers are largely eliminating *Nothofagus* forest regeneration and facilitating the invasion of introduced plant species along river corridors altered by beaver activities. In other words, beavers are engineering the ecosystem to create their own habitat, but in this case they are eliminating the habitat of native species.

Where Beavers Belong

At the North St. Vrain beaver meadow, many of the natives are still here. The grizzly bears and wolves described only a century ago by Enos Mills are gone. The beavers remain, but they hang on by a proverbial toehold as a result of severely limited habitat and continued trapping outside of the national park boundaries, as well as competition by large numbers of elk and moose no longer culled by predators. In former beaver meadows to the north and south, the beavers are gone.

Standing beside a small dam on a side channel of North St. Vrain Creek, I can see across to a recently breached pond floored with dark brown muck. The steep banks along the drained pond reveal layers of organic-rich soil and beaver-gnawed branches above the cobbles that mark an abandoned course of the creek. Off to one side, a rust-colored seep of water stains the pond floor, leaving a hint of sulfur in the air. I snowshoe across the frozen mud to the center of the pond, where I have a clear view back up the valley to the higher peaks, mostly covered in snow but for a few darker patches of rock on the northeastern face. A rampart of tall storm clouds rises to the west beyond the peaks. The track of the glacier upstream is obvious as a long, winding trough gouged into the uplands. Once the glaciers retreated, the creek and the beavers became partners in shaping this riverscape.

I continue downstream and come across the tracks of a beaver—long, slender, web-toed hind feet, and long-toed front feet, with the drag mark of the tail broad as my snowshoe tracks. The tracks emerge from the pond at a steep, snowy bank immediately above a cache of branches pale where the bark has been stripped, just visible through the water where the beaver has broken the ice. The beaver wandered about the bank a little, leaving a pile of soft, greenish-brown beaver poop that appears to be mostly bits of plants. Like other herbivorous animals, beaver can re-ingest their own feces and extract a few more nutrients from their food. I knew that the beavers could survive the September flood in a meadow this broad and varied, but I still find it heartening to see the evidence of their continuing presence. It is difficult to overcome the human perception of a flood as a life-threatening disaster.

I turn back upstream, noting stashes of stripped branches in the side channels all around, but no sign of an active lodge. A mountain chickadee flits lightly from branch to branch above a narrow side channel before moving on. I try not to collapse too many of the hollow pockets that I know shelter smaller creatures, but every

An abandoned beaver dam that North St. Vrain Creek, at right, has cut through. The dam is a mixture of beaver-gnawed branches, finer organic matter, and sediment. A color version of this figure is included in the insert section.

so often one of my snowshoes plunges deeper among the willow stems. I find coyote tracks on top of the tracks I made on my way downstream: the meadow comes more alive the longer I continue.

The strident calls of a Steller's jay attract my attention upward. I see a flash of azure feathers as the bird flies up the valley side slope where aspens stand out as lighter patches of gray trunks and branches among the dark green conifers. As I return to my entry point, I hear the soft plashing of the moose's footsteps as it walks up the creek.

March

Water Superheroes

The Meadow in March

By mid-March, daytime temperatures above freezing have left muddy puddles all over the unpaved road that runs above and beside the beaver meadow. This road extends to the national park trailhead farther upstream but is now closed for winter.

I enter the beaver meadow on a lightly overcast day that is windy, as I expect March to be. Lack of recent snowfall and warm temperatures have caused the snow-pack to shrink down, and I no longer break through into hidden pockets of air around the base of the bushy willows. I do break through the ice on my snowshoes, sinking in a slow motion that allows me to scramble and keep my feet dry . . . mostly. I sink in above the ankle at one point and the resulting icy ache makes me appreciate the ability of beavers to stay warm.

The snow covering the higher peaks and the adjacent lateral moraines appears about the same, but numerous spots of bare ground have appeared along the creek banks. The remaining snow resembles a blanket draped over the undulating, grassy ground rather than an integral part of the landscape. I stand on the snowbank at the downstream end of one of the larger beaver ponds. The dam merges into a vegetated berm and appears to be intact, but I can hear water flowing swiftly somewhere be-neath the snow. Most puzzling is that I can't see where the water is going: the nearest downstream standing water has no apparent inflow or current. Mysterious, intricate plumbing surrounds me. The beaver meadow is on the move, flowing and changing, preparing for the season of birth and growth.

Standing water is noticeably more abundant than a month ago. Interspersed among the ice and snow are big puddles and little ponds, some connected and draining, others isolated and still. The still pond waters have a shallow covering of meltwater underlain by ice with large, irregularly shaped air pockets trapped in the upper layer. These I can easily break with the tip of my ski pole. Thousands of tiny bubbles deeper in the ice look milky. Clumps of pine needles, twigs, and stems of

Snow mounded along a side channel of North St. Vrain Creek. The glacial lateral moraine appears as a forested ridge in the background. A color version of this figure is included in the insert section.

grass frozen into the ice create a layered time capsule of events in the pond since the ice began to form in autumn. These little ponds are incubators for the biogeochemical reactions that will soon power an outpouring of life as summer comes to the beaver meadow.

For now, the sand bars exposed along the creek are frozen only a couple of inches below the surface. The mainstem flows freely, still clear and free of the fine sediment that will be released by snowmelt on the uplands, but already higher with the start of the melt. The cobble bars reworked by last September's flood are pale orange among the darker orange-brown stains left by iron oxide seeping from the ground and tannic acid carried by the creek.

Two weathered beaver lodges show no signs of recent activity. The spot where I found beaver tracks in February is no longer tracked, although aged beaver poop remains on top of the snow. Bird calls become noticeable during intervals in the wind. Otherwise, the only sounds beyond the flowing creek are the rustling hiss of wind through grasses or the louder shhh-ing of wind among the willow branches and pine boughs. I know, however, that the relative silence conceals plenty of beaver activity, as well as the consolidation of the snowpack that will give rise to the great annual rush of water from this mountainous catchment..

Fire, Flood, and Drought

Scientists across the Northern Hemisphere have quantified the effects of beaver dams and ponds on streams, but few have been fortunate enough to witness the type of flash flood that Enos Mills described in his 1913 book *In Beaver World*:

> The easy rain of two days ended in a heavy downpour—a deluge upon the mountain-side a mile or so upstream. There was almost nothing on this mountain either to absorb or delay the excess of water which was speedily shed into the stream below. Flooding down the stream's channel above the beaver pond, came a roaring avalanche of water, or water-slide, with a rubbish-filled front that was five or six feet high. This expanded as it rolled into the pond, and swept far out on the sides, while the water-front, greatly lowered, rushed over the dam. A half a dozen ponds immediately below sufficed so to check the speed of this water and so greatly to reduce its volume that as it poured over the last dam of this colony it was no longer a flood. (p. 215)

Part of the interest in this historical description is that Mills describes the effect of a string of dams and ponds. This scenario is now much rarer: only a single pond or a small number are likely to be present along streams examined by scientists today.

Heavy rains can certainly cause a beaver dam to fail, resulting in an abrupt, short-lived flood known as an outburst flood. An outburst flood can be particularly forceful in changing stream channels and damaging infrastructure because the flood occurs so rapidly and unpredictably and can have a very high peak flow immediately downstream from the dam. David Butler, in particular, has documented outburst floods from beaver dams in the southeastern United States, Montana, and elsewhere in North America. The failure of multiple dams in sequence, however, is rare, suggesting that where healthy beaver populations maintain multiple dams along a stream, the chances of a damaging outburst flood are much lower.

Water storage in streams with multiple beaver dams is also likely to be more effective than in streams with a single dam and pond. Working in the Adirondack Mountains of New York, Douglas Burns and Jeff McDonnell compared stream flow in a watershed with a three-acre beaver pond to an otherwise similar watershed with no beavers. The beaver pond affected downstream movement of water after precipitation via evaporation from the pond and mixing of pond and stream water, but the single pond did not retain a large volume of water during periods when the entire watershed was shedding water, such as during snowmelt. On the other hand, studies by Cherie Westbrook and others in the upper part of the Colorado River on the west side of Rocky Mountain National Park demonstrated that a series of beaver dams significantly influenced the movement of water during a 10-year flood.

The dams greatly increased the depth and spatial extent of overbank flow, as well as the time during which the floodplain was inundated. Compared to portions of the river without beaver dams, the dammed reach had a higher groundwater level during times of both high and low stream flows. The dams effectively kept the valley bottom from drying as much during the hot summer months.

Ming-ko Woo and James Waddington used numerical modeling of a stream in southeastern Alaska to infer that a single dam can reduce peak stream flows by 5%, but a series of five consecutive dams and ponds can reduce peak flows during a two-year flood by 14%. These numbers represent minimum estimates because the scientists assumed that the ponds were full and water flowing over each dam immediately returned to the main channel rather than soaking into the valley-bottom sediments or spreading across the floodplain. At a site where beavers were recently reintroduced in Belgium, Jan Nyssen and his colleagues found that a series of six beaver dams on the Chevral River lowered peak flows and slowed the downstream movement of floods by a day, as well as increasing the low flows.

An important thing to keep in mind when reading these studies focused on single beaver ponds or relatively small valleys is the potential cumulative effect of thousands of beaver ponds and meadows across a region. Few scientists have systematically inventoried beaver activity across a wide region, but studies during the 1980s and 1990s indicate that beaver ponds can occupy from much less than 1% to over 6% of contemporary landscapes within the contiguous United States. These low numbers likely reflect the long history of landscape modification and beaver removal.

A more recent effort by Alasdair Morrison and others used aerial imagery and ground surveys to examine an area of 3,050 square miles in the foothills and mountains of the Canadian Rockies around Banff, British Columbia. Scientists estimate that wetlands make up only 2% of the land area in the Rocky Mountains as a whole. Of the wetlands identified by Morrison and his colleagues in the area around Banff, beavers created nearly half (43%) of those in the mountains and about a quarter (26%) of the wetlands in the foothills, where human disturbance limited beavers. In protected areas such as parks, beavers created 60% of the wetlands. For the hundreds of wetlands that the scientists studied, those formed by beavers had 10 times as much open water as wetlands formed in the absence of beaver. Not all of the wetlands were along rivers, either. In the absence of channels, beavers can dredge canals and dam seeps and springs fed by groundwater. In other study areas, scientists have documented beaver dams across lake outlets, ditches and culverts, bogs, fens, deciduous swamps, and tidal forested wetlands. Across the entire area studied by Morrison's team, wetlands covered just over 28 square miles, or less than 1% of the total area. Think about the effect of removing all of the beavers and potentially drying up 30% to 60% of this already-small portion of the landscape that is capable of storing water, providing habitat for diverse plants and animals, and releasing the stored water gradually downstream.

Beaver engineering of water movement can be especially important during times of water stress. Working in Elk Island National Park in east-central Alberta, Glynnis Hood and Suzanne Bayley demonstrated how beavers can buffer the effects of regional drought. The presence of beaver dams in their study area created a nine-fold increase in the area of open water during both wet and dry years when compared to historical periods during which beaver were absent from the same sites. Analogous results come from studies in northern California, Missouri, Oregon, Texas, Utah, New Mexico, Colorado, British Columbia, and Germany.

A small stream can remain perennial if the amount of sediment underneath the stream is sufficient to store the water infiltrating from beaver ponds. The sediment acts like a bank account for water. The ponds "deposit" water in the underground bank by gaining in volume during wet periods and slowly leaking surface water into the ground. This recharges the groundwater, which can slowly feed the stream and keep it flowing during dry periods.

In each of the studies mentioned above, beaver dams create longitudinal disconnectivity, slowing the downstream movement of stream water and allowing the water to be stored at the surface in ponds, as well as infiltrating into the ground from ponds, overbank flow across the floodplain, and flow within secondary channels. The stream can, in effect, withdraw these groundwater deposits during periods when not much precipitation is falling. This effect was also demonstrated in a study by Ming-Ko Woo and James Waddington in which they compared the water balance of catchments in the subarctic portion of North Ontario that did and did not include beaver dams. Although catchments with beaver dams lost more water to evaporation, this loss was less important than reduced peak flows and increased water storage within the dammed catchment.

Scientists have recognized for at least three decades that beaver dams increase flow from the channel into the streambed and underlying sediment. David White described the phenomenon along Michigan's Maple River in 1990. More recent research has quantified the effect of a single beaver dam on this surface–subsurface exchange. However, we still know little of the mechanisms by which an entire beaver meadow with numerous dams fills the bank account of valley-bottom sediments with water during periods of high flow and then gradually releases the water during periods of low stream flow.

Beavers are not immune to the vagaries of climate. Interpreting sediments deposited over the past 11,500 years along valley bottoms in Yellowstone and Grand Teton National Parks in Wyoming, Lyman Persico and Grant Meyer found that beavers left smaller streams when these streams stopped flowing during warmer, drier episodes of climate. During this long time span, three periods of cooler, wetter climate—4,200 to 2,500 years ago, 1,700 to 1,000 years ago, and the past 500 years—have been good times for beavers. During these times, which total approximately 2,900 years, beaver ponds trapped between 60% and 76% of the total thickness of sediment deposited during the 11,500-year time span.

On slightly larger streams, beavers are quite capable of surviving through periods of drought, as demonstrated by the study from Alberta. Glynnis Hood and Suzanne Bayley observed beavers digging canals into the bottom of their shrinking ponds during dry times, deepening the pond and trapping any remaining water. In her book *The Beaver Manifesto*, Hood describes how the beaver canals collected available water at the entrance to their lodges and along access routes to favored foraging areas. This was not a simple task: in some cases, the beavers spent more time digging the canals with their front paws than collecting food for their winter food caches. At the height of the drought, ponds that still contained water were mostly occupied by beavers. The reward of the most skillful beavers was survival. Shallow ponds froze to the bottom. Hood found that when the beavers trapped by these frozen ponds chewed through the sides of their lodges to escape, coyotes were waiting to eat them. A study of moderate-sized streams with and without beaver dams in Indiana revealed that the streams with dams can retain water at least half again as long as those without dams.

Within a beaver meadow, individual ponds can fill or be abandoned by the beavers, even while newly built dams are ponding water elsewhere within the meadow. With time, the beavers create a mosaic of differently aged ponds and different habitat for plants and animals. Some ponds can be quite persistent, however. David Butler and George Malanson described ponds that were more than 40 years old in Montana's Glacier National Park. These beaver ponds were built on the deltas of glacial lakes and were particularly large and stable. Ponds covering 0.4 to 2.5 acres stored anywhere from 1,150 to 8,310 cubic yards of sediment per pond. Other beaver ponds within Glacier had annual rates of sediment accumulation much higher than other portions of the landscape, and the oldest ponds contained the greatest amount of organic matter, effectively serving as a sink for organic carbon.

Wildfires are presumably one of the greatest stresses to a beaver colony, not only because of the heat and smoke of the fire itself but because of the loss of living vegetation that beavers eat and the way in which the denuded landscape sheds water and sediment, leading to floods and debris flows after the fire. Enos Mills described a beaver meadow that withstood a fire in Rocky Mountain National Park, creating a verdant patch in an otherwise charred landscape.

I observed a very different scenario following an autumn 2012 wildfire in the national park. An illegal campfire near Fern Lake started a wildfire that burned for months at low to moderate levels before winter snows finally extinguished the flames. The upstream portion of Moraine Park along the Big Thompson River was one of the areas that burned. Four decades earlier, when beavers maintained wet meadows in Moraine Park, this area likely would not have burned. Overgrazing of riparian willows by elk had reduced the food supply for the beavers, causing beavers to abandon the site. As the beaver dams fell into disrepair, flow concentrated in a single channel rather than spreading across the valley bottom. The beaver meadow dried into an elk grassland that burned readily during the fire.

When beavers and their dams go, a cascade of secondary effects occurs. Kim Green and Cherie Westbrook used aerial photographs to assess changes along Sandown Creek in British Columbia after 18 beaver dams were removed from the creek during the late 1980s. In the first 16 years following removal of the dams, the riparian area changed from 69% sedges and grasses and 9% beaver ponds to 90% shrubs and trees. The creek changed from multiple channels that branched and rejoined to a single channel. Green and Westbrook estimated that this increased the average velocity of stream flow by five times. The faster water enlarged the stream channel and eroded pond sediments. Working in Montana, Rebekah Levine and Grant Meyer also found that much of the sediment stored in a beaver pond can be eroded and carried downstream when the dam is breached. Not all of the sediment, however, is eroded. The remaining fine-textured soils are cohesive and resist bank erosion, helping to form meandering channels and floodplains.

As I watch a March snowstorm alter the appearance of the North St. Vrain beaver meadow, I imagine the landscape of Rocky Mountain National Park 200 years ago, before commercial trappers began to remove beavers. As someone who studies rivers, I'm sure that what would have struck me first was the much greater wetness of the valley bottoms. In the wider valley segments just upstream from each glacier terminal moraine, sequential beaver dams would have created a stepped valley of ponds separated by short drops. Dense stands of willow and aspen likely hid the abundant ponds unless the landscape was viewed from the top of a nearby ridge. Waterfowl probably were more abundant, along with the many other animals that like wetlands. The contrast between lighter green aspen and willow foliage and darker green conifers would clearly distinguish the edges of the valley bottom from the adjacent, drier hillslopes. The sight of a burned valley bottom would be rare or nonexistent.

Water Superheroes

Water management decisions do not always reflect the abundant scientific evidence that beaver dams and ponds buffer the effects of fire, flood, and drought. Although I had not yet begun to actively conduct research on beavers and their effects on streams, I was struck by the absurdity of a directive issued by the Colorado State Engineer's Office during a severe drought in 2003. The state engineer suggested removing beaver dams across public and private lands in order to increase stream flow. Enos Mills knew better a century ago.

In her 2011 book *The Beaver Manifesto*, Canadian ecologist Glynnis Hood called for a "water superhero" to help us as the climate warms and, in many regions, grows drier. As she noted, perhaps the beaver is just the ally we need. I think of this as I compare the effects of the September 2013 flood at the North St. Vrain Creek beaver meadow versus the effects at the Fish Creek area of Estes Park.

Fish Creek flows from Lily Lake, about six miles north of the North St. Vrain beaver meadow. The creek initially descends steeply from the lake, then flows at a much gentler gradient along the southern end of Estes Park before joining the Big Thompson River in the artificial reservoir of Lake Estes. Suburban housing, a ranch with horse pastures, and the Fish Creek Road narrowly constrain the creek along most of its course through Estes Park. Beavers built a dam and lodge along Fish Creek in 2011, causing consternation in some of their human neighbors and delight in others. Concerns arose from the potential for the beaver pond to flood portions of the road and adjacent properties and from the beavers' very efficient harvest of aspens growing along the creek. Delight arose because . . . well, because they were beavers and they were not camera-shy when people stopped along the road to photograph them.

As I have described already, the 2013 flood had almost no noticeable effect on the North St. Vrain beaver meadow, and the beaver meadow probably helped to attenuate the floodwaters and prevent damage to the road and culvert at the downstream end of the meadow. Thick, dark soils exposed in the banks along Fish Creek indicate that the portion of the creek upstream from the 2011 beaver dam had been a beaver meadow historically, although it was an overgrazed pasture with few willows when the beavers moved in downstream. The 2013 flood caused extensive damage along Fish Creek, eroding the channel (and the road) where the creek flows through the pasture and suburban areas and then dumping a large volume of sediment where the creek enters Lake Estes. The beaver dam was completely destroyed and the flood cut a deep trench into the Fish Creek channel where the dam had been. By May 2013 there were signs of new beaver activity at the former dam site, so the beavers survived, but they have their work cut out for them.

The effects of the 2013 flood along Fish Creek likely would have been very different if an active beaver meadow with abundant willow thickets and numerous dams had been present along the portion of the creek within Estes Park. We have a water superhero in beavers, but we need to get out of their way. The 2013 flood caused extensive damage along rivers throughout the Colorado Front Range because of the infrastructure and people who had encroached on the floodplain since the last major flood—which, for the Big Thompson River, was less than 40 years ago, in 1976. After the 2013 flood, everyone talked about "cleaning up" the rivers, which meant channelizing them and removing all of the sediment and wood that the flood had deposited in channels and across floodplains. Cleaning up rivers should more properly refer to removing the infrastructure—roads, pipelines, houses, and other structures—that has been inappropriately built within river floodplains. During the flood of 2013, every little creek and major river in the region reoccupied its historic floodplain—the nearly flat surfaces adjacent to each channel that have been deposited and reworked by flowing water over a period of centuries. These temptingly flat sites with a scenic river view are there because they are the product of repeated river erosion and deposition, and it defies common sense to expect that river erosion and deposition will cease simply because people choose to occupy these surfaces.

The words used in connection with floods are worth examining. Disaster: phrases such as "natural disaster" or "flood disaster" were widely used after the 2013 flood. A disaster is a sudden or great misfortune, a calamity. Floods are not disasters for river ecosystems. Floods are disturbances, which ecologists define as a change in average environmental conditions that causes a pronounced change in an ecosystem. Disturbances can be highly beneficial to ecosystems by creating opportunities for pioneering species that move into an area immediately following disturbance. This increases the species diversity of the ecosystem, assuming that some portions of the ecosystem were not completely altered and continue to support late-successional species.

Floods become disasters for people because of how we interpret probability and how our actions affect hazard and risk. Probability is the degree of likelihood that something, such as a flood, will occur. Hazard is the risk of loss or harm. Risk is exposure to the possibility of loss. We turn floods into disasters by gambling on the probability that a flood will not occur, risking ourselves and our property, and

A cutbank about four feet tall along Fish Creek after the September 2013 flood. This site, upstream from the dam that beavers built along the creek in 2011, has the thick, dark, organic-rich soils characteristic of a beaver meadow, although at present the site is a dry pasture. A color version of this figure is included in the insert section.

increasing our hazards by occupying floodplains. When you inadvertently commit a minor crime such as a traffic violation, the standard response is that ignorance of the law is no excuse. The stakes are much higher and judgment in the form of a flood is much more implacable, but the principle is the same: ignorance of the processes that create and maintain channels and floodplains is no excuse.

Beavers can be water superheroes in many respects. They can help to sustain flows of clear, clean water. They can attenuate the effects of fires, floods, and droughts on the landscape. But beavers can only fulfill these vital functions if we collectively recognize the importance of their activities and keep ourselves and our stuff out of the valley bottoms where rivers create floodplains and beavers create meadows.

I think of this as I stand on the bank of North St. Vrain Creek and listen to the voice of the creek, which is noticeably louder than it has been since January. Water trickles steadily down the side channels, where green patches of filamentous algae mark the points at which water wells up into the channel from the subsurface. I scoop up a handful of the sand and gravel beneath the algae and watch aquatic insect larvae wriggling vigorously across my hand before returning sediment and bugs to the water. I sense the vernal gathering of energy in the beaver meadow. Despite the newly fallen snow, the trend now is toward snowmelt. Like the warming air and the swelling, greening plants, the rising stream flow creates a sense of rapid change and expectancy. The most energetic season of the river's year is coming, but the beavers and their dams are ready for it.

April

Six Degrees of Connectivity

By late April, the snow is gone from the beaver meadow. The promises of March are starting to be fulfilled: insects are on the wing, some of the willows have furry catkins along their branches, and fish jump from the quiet waters of the beaver ponds. I can no longer easily get around the beaver meadows on foot unless I wear chest waders. The sound of the beaver meadow in March was primarily wind. By April, the sound is primarily moving water. The water gurgles, shushes, and whispers. In another month it will roar with the melting snows. Another three miles up the creek valley and 1,500 feet higher, one of my long-term study sites still lies under 6 feet of snow, but in the meadow I see only one patch of tenacious snow-ice in the deep shade beneath a spruce along the northern edge of the meadow. I know that snow will still fall here during late spring storms, but it will melt quickly. March felt on the cusp, as if it could as easily tip toward winter or spring. Late April is definitely spring headed toward summer.

The beaver meadow remains a riverscape more brown and tan than green. The willows are still leafless, although some of the branch tips are turning pale yellow-green and others seem to be taking on a more vivid orange hue. I can see the leaf buds starting to swell. The grass has just begun to grow in dark green tips steadily forcing their way through the thick mat of last year's dead stems. Clusters of new leaves on low-growing wintergreen are the only other sign of green outside of the channels. Some of the smaller side channels are thick with emerald green algae undulating slowly in the current. A stonefly lands on my hand. Its slender, dark gray body seems surprisingly delicate for a creature that has hatched into the vagaries of April air, with its potential for blasting winds and sudden snow squalls.

I look around, trying to pinpoint what quality is giving me such a feeling of expectancy. The western chorus frogs are calling steadily. I smell pine sap on the dry terrace along the beaver meadow, where burrowing rodents have left linear mounds of soil displaced by their digging. Down in the meadow, each footstep stirs up a smell of wet, fecund muck. The sun is farther north and the air feels warmer and softer. The light appears brighter and the color palette of the meadow broader than a month ago, although there is no question that even December sunshine on snow beneath a blue mountain sky can form a stunningly bright color palette in Colorado. I feel more expansive and optimistic in the warmth and light, ready to see beaver kits emerging from their den. I probably have more than a month to wait yet.

Meanwhile, the kits are undoubtedly learning important lessons from their family members, as well as being waited on paw and paw. Although beaver kits are born with eyes open, fully furred, and able to swim, the world outside the lodge remains an extremely dangerous place for them for at least one simple reason: baby beavers are too buoyant to dive. All of the openings of a beaver lodge are below the water surface, so that only a diving animal can make use of them. A baby that ventured out on its own would be unable to return to the lodge. So, the babies remain well attended. Every member of the family helps to deliver fresh greens, regularly change the bedding of woodchips or grass, and babysit. Hope Ryden writes in *Lily Pond* of the momentous day when the baby beavers become old enough to dive underwater and are ushered from the lodge by attendant parents and other colony members: "An entourage of adults and yearlings surrounds the youngsters and provides them handy backs on which to hitch rides in the event that they grow

One of the main beaver ponds, with a lodge in the center. The ice and snow are gone and the pond is open, but the willows have not yet begun to leaf out. A color version of this figure is included in the insert section.

tired" (p. 60). Ryden noted that although the young beavers are able to swim and dive well after a few weeks, they stick close to the adult escorting them and wail if the adult starts off without them. If there is a wailing baby anywhere about on this April day, I cannot hear it above the calls of the frogs and the myriad sounds of water on the move.

A Rolling Sand Grain Makes No Soil

Removal of the snow cover reveals a large, drained pond on the north side of the beaver meadow. The freshly exposed pond sediment has not yet been colonized by any vegetation, although that will start to change as the growing season advances. For now, the pond sediment, fissured into cracks where it is drying in the sun, simply represents a mass of sand, silt, and clay that did not go downstream.

Much has been made of the ability of beaver dams and beaver ponds to trap and store sediment. Sediment moves downstream within a river in proportion to the supply of sediment and the energy of the flowing water. Energy results in velocity. Flowing water obstructed by a beaver dam or spreading into the still waters of a beaver pond loses velocity and in the process loses some of its ability to carry sediment.

Enos Mills described these effects in his 1913 book, inferring the connections between sediment accumulating in contemporary beaver ponds, the filling and abandonment of these ponds, and subsequent colonization of the pond sediment by vegetation. Once you look for it, the connection is not hard to make. Driving the roads through Estes Park, plenty of stream banks cut into now-dry grasslands reveal the thick, black soils that in this region only form beneath the waters of a beaver pond.

Rocky Mountain National Park and the mountain ranges and prairie surrounding it are semiarid. Just enough rain and snow fall to maintain vegetation that covers most of the ground surface, but the region is on the edge of being arid. Multiple times during the past 10,000 years, that tipping point has been crossed in the Great Plains at the eastern base of the mountains, and what is now shortgrass prairie has been shifting sand dunes. This happened most recently during the Great Dust Bowl of the 1930s and during a lesser-known dust bowl that affected parts of eastern Colorado in the 1950s, when the vegetation loss, soil erosion, and dust were worse than the 1930s episode.

Even the mountains, with their abundant snowfall relative to the plains, are not wet enough to create the thick, rich soils common east of the hundredth meridian or along the wet coastal ranges of the Pacific Northwest. The lack of moisture has at least two effects. First, dryness keeps the bedrock from breaking down into sediment. Bedrock weathers into sediment through physical processes such as freezing and thawing that break the cohesive rock into fragments and through chemical processes when materials dissolved in water exchange elements and ions with the minerals in the bedrock. These physical and chemical processes occur in the Colorado Front Range, but slowly. The resulting soil tends to be thin—only a few inches deep on most hill slopes—and composed of sand- to gravel-size bits of granite rather than thick loam. Overlying the granitic base is a layer of litter and duff composed of plant parts such as twigs, leaves, and blades of grass shed from whatever vegetation grows on the site. These bits of plant detritus gradually decay from litter, in which the original plant parts are still discernable, to duff, which is composed of an unidentifiable mass of organic material. Much of the nutrient base that support living plants is contained within this upper layer, but the upper layer is not especially thick—typically five inches or less. This is the second primary effect of limited moisture on soil development: there is not enough water to support a dense vegetation community, so limited amounts of plant detritus drop onto the ground and contribute to the soil.

Two processes help to develop thicker, richer soils at some sites in the Front Range. The first is windblown dust or, more specifically, windblown particles of silt and clay that settle onto the ground. Windblown dust is currently receiving a lot of attention in the western United States because the amount of dust being deposited on mountain snowpacks has increased tremendously in recent decades, with strong effects on the rate at which snowpacks melt in spring. Dust has been falling from the sky in the western United States for thousands of years, however, because the region has been dry for thousands of years.

Silt and clay represent a sort of end product of bedrock weathering: prolonged and intense weathering of a type that does not occur in the Front Range is necessary to produce silt and clay. Silt and clay also help to stabilize soil, increasing the cohesion of the soil particles and thus limiting erosion, and also increasing the ability of the soil to retain moisture and support vegetation. In other words, silt and clay help to form a thicker, richer soil.

Another process that helps form a thick, rich soil is to pond water along a stream valley. The sediment and organic matter coming down the stream can accumulate in the pond, and plants such as sedges, rushes, and algae that grow in the pond add more organic matter when they die and decompose. Poke a stick into the thick, black mud at the edge of a beaver pond and inhale the swamp smell that results: that is good soil in the making. Only marshes, fens, and other wetlands create this type of soil in the semiarid Colorado Front Range.

The nature of the thin, rocky soil covering much of the watershed of North St. Vrain Creek was abundantly clear after the September 2013 flood. Buttonrock Dam, about 12 miles downstream from the beaver meadow, impounds the waters of North St. Vrain Creek in a reservoir. The heavy rainfall associated with the September flood caused numerous landslides along the steep valley walls between the beaver meadow and the reservoir, and the swollen flow of the creek eroded stream banks and floodplains. More than 350,000 cubic yards of sediment were dumped into the upstream end of the reservoir as the floodwaters lost velocity. This sediment is mostly sand- to boulder-size material, not silt and clay.

To Make a Meadow, It Takes a Beaver and One Dam

Scientists began to write about the sediment trapped behind numerous beaver dams nearly a century ago. A 1938 paper by Ruedemann and Schoonmaker published in the research journal *Science* noted that older settlers around parts of New York used the phrase "beaver meadow" to describe the broad, flat alluvial valleys that resulted from beaver ponds gradually filling with sediment. The authors described the mosaic of active and abandoned beaver ponds common in these meadows and wrote of how continued accumulation of sediment could with time create a thick fill of soil across a valley bottom. Other scientists have been fascinated with this process ever since.

In 1942, Ronald Ives described the interactions between beaver dams, filling ponds, and vegetation colonization over decades to centuries and how these interactions resulted in storage of sediment and development of valley floors in Rocky Mountain National Park. In 2011, Cherie Westbrook, David Cooper, and Bruce Baker systematically examined these interactions in detail along the Upper Colorado River valley on the western side of the national park. They measured how one newly built dam 5.5 feet tall triggered overbank flooding. In some places, the deep waters of snowmelt floods drowned existing vegetation. In other places, the floodwaters deposited nutrient-rich sediment on the floodplain or eroded the existing sediment. High stream flows subsequently breached the dam and the site dried out, but the remaining bare sediment was quickly colonized by diverse plants, including willow, aspen, *Carex* sedges, and grasses. The resulting vegetation patches created diverse habitat for insects, birds, and other animals.

These processes are not unique to Rocky Mountain National Park. David Butler and George Malanson painstakingly measured sediment stored in beaver ponds in Glacier National Park, Montana. Younger ponds store anywhere from 12 to 216 cubic yards of sediment, but older ponds store 100 to 350 cubic yards of sediment. Other scientists measured 45 to 8,500 cubic yards of sediment in beaver ponds along boreal streams in eastern Canada and 7,900 to 20,000 cubic yards of sediment behind individual beaver dams in Quebec. Assuming an average rate of sediment accumulation based on the size and age of a beaver pond, individual studies report sedimentation rates ranging from less than 0.2 inches per year in a headwater beaver pond in central Ontario to nearly 11 inches per year in a pond in the Rockies within Montana.

The size of a beaver dam or pond is not necessarily a good predictor of how much sediment the pond contains. The amount of sediment coming down the stream exerts an important influence on sediment storage. Factors such as upstream geology and the land use history of the watershed control the amount of sediment in transport. Disturbances such as wildfires or floods can result in at least temporarily greater sediment movement downstream. As ponds fill, emergent vegetation rooted in the bottom sediment but extending above the water surface grows toward the center of the pond, helping to trap more sediment and accelerate the filling. Each beaver dam along a stream can trap sediment, so the number of dams upstream from a particular site affects the amount of sediment reaching that site. The stability of upstream dams and dams at the site is also important. Every time a beaver dam fails, some of the sediment stored behind the dam is likely to be eroded by the newly accelerated flowing water. Finally, the shape of the valley bottom influences the amount of sediment stored upstream from a beaver dam. A steep, narrow valley simply does not have much space to accumulate sediment.

People trying to restore streams have taken advantage of the ability of beaver dams and ponds to reduce the velocity of stream flow and trap sediment. One restoration project focused on Currant Creek in Wyoming, a stream that had cut deeply into the valley-bottom sediment. In 1980, 33 tons of silt a day traveled down the creek, which continued to erode its nearly vertical banks. Some aspen trees remained along the valley bottom, providing a food source for beavers that were introduced to the site as part of the restoration effort. The beavers did what comes naturally, building dams that trapped sediment and ponded water. Some of the ponded water soaked into the stream banks and flowed across the valley bottom, raising the water table. Willows began to germinate along the banks, stabilizing the banks against further erosion. Within two years, the sediment load dropped by 90% to four tons per day. Joe Wheaton, William Macfarlane, and their colleagues at Utah State University have developed numerical models to study the distribution of habitat suitable for beavers and the number of beavers a valley bottom can support, and beavers are being used for stream rehabilitation at sites in Washington, Wyoming, and elsewhere.

Beavers Versus Glaciers

One of the scientific debates generated by Ronald Ives's 1942 work was the relative importance of beavers and glaciers in creating the broad, flat valley segments scattered along otherwise steep, narrow river corridors in the Rocky Mountains. Working in Yellowstone National Park, Lyman Persico and Grant Meyer argued that individual beaver dams are typically only a few feet tall, so that the sediment trapped behind each dam is of limited thickness. Persico and Meyer painstakingly measured sediment accumulations exposed along numerous stream banks and concluded that the thickness of sediment associated with beaver dams was a minor contributor relative to the sediment piling up in alluvial fans after wildfires or dropped by melting glaciers.

Working with graduate student Natalie Kramer and my colleague Dennis Harry, I measured the total thickness of sediment accumulated in a valley bottom in Rocky Mountain National Park since the retreat of the Pleistocene glaciers and the proportion of that sediment associated with beaver dams. We chose Upper Beaver Meadows, a site along Beaver Brook, to the north of the North St. Vrain beaver meadow. Despite the name, there are no beavers in this meadow today. Beavers were present only a few decades ago, but their food supply was largely eliminated by intensive elk grazing of riparian willows after elk predators such as wolves had been hunted to extinction in the national park.

We used a geophysical technique known as ground-penetrating radar to image the subsurface layers at Upper Beaver Meadows. As the name implies, radar waves are transmitted into the ground, where they are reflected back to the surface in a manner that depends on the grain size, moisture content, and horizontal layering of subsurface sediment. Basically, we created images that resemble static on a television screen. Interruptions in pattern within these images signal a change in the subsurface material. The patterns themselves reveal only limited information: interpretation relies on calibrating the patterns by being able to recognize a pattern characteristic of a particular type of buried feature. Upper Beaver Meadows was abandoned recently enough that there are beaver dams and pond sediments at the surface. These features are covered in grasses, but they are easy to spot because the old dams form linear mounds across the valley and beaver-gnawed wood sometimes protrudes from the mound. We used these shallow dams and pond fills to understand the pattern created by more deeply buried beaver dams and ponds. Using this approach, we estimated that about half of the sediment that has accumulated in the valley bottom since the Pleistocene glaciers retreated is stored by old beaver dams and ponds.

This seems to contradict the interpretations of Persico and Meyer, but the scenarios are not mutually exclusive. The relative importance of beaver-accumulated sediment strongly reflects the details of each valley segment. Working with graduate student Zan Rubin and colleague Sara Rathburn, Dennis and I also used ground-penetrating radar to image a portion of the Upper Colorado River valley on the western side of Rocky Mountain National Park. This portion of the park includes the former gold mining site known as Lulu City and a segment of river valley known as Little Yellowstone that lies below Specimen Mountain. The names reflect the mineralization of the bedrock in this part of the national park. Millions of years ago, superheated water rich in dissolved metals was forced upward into cracks in the bedrock, where it cooled and deposited veins of ore. This process weakened the bedrock, making it more susceptible to weathering and erosion. Debris flows are more common along this portion of the Upper Colorado River valley than along the eastern valleys such as Beaver Brook. Although we found some beaver-ponded sediment in the valley-bottom sites along the Upper Colorado, much more of the total sediment fill resulted from debris flows, analogous to the fire- and debris-flow–prone environment in Yellowstone.

On the eastern side of Rocky Mountain National Park, the most extensive beaver meadows form just upstream from the terminal moraine left by each valley glacier. By at least temporarily blocking the entire valley with a mound of sediment perpendicular to the downstream movement of water and sediment being released by the melting glacier, the terminal moraine started the formation of a broad, flat valley bottom. After the glacier retreated and vegetation colonized the valley bottom, this formed an ideal environment for beavers to build multiple dams and create the mosaic of streams, ponds, dams, willow thickets, and wet meadows that compose a beaver meadow. What the glaciers started, the beavers perfected.

Only Connect . . . or Disconnect

There are fashions in science, as in any human endeavor. Someone comes up with a good idea that helps to organize many observations and provides a framework for asking relevant questions. At present, connectivity is in fashion. Ecologists, hydrologists, geomorphologists—those who study natural environments—try to understand, map, and measure connectivity within environments. How do water, sediment, and nutrients such as nitrogen move through river networks (riverine connectivity)? How do organisms move between suitable habitat patches on a landscape (functional landscape connectivity)? How does sediment move from the hillslope where it is created by weathering bedrock to the channels where it is transported (sediment connectivity)?

Riverine connectivity has spatial and temporal dimensions. Materials and organisms do not just move downstream within channels. They can also move upstream (think of spawning salmon or migrating eels) or between the channel and the adjacent floodplain (fish spawning on the floodplain or sediment being deposited by water moving slowly across the floodplain). There are exchanges of matter and organisms between the channel and the subsurface, such as water seeping into or upwelling from the streambed or larval aquatic insects burrowing into the bed sediments. And there are exchanges between the channel and the atmosphere, ranging from precipitation falling on the river to hatching insects emerging from the channel as winged adults.

I like to think of river channels as having six degrees of connectivity—with the atmosphere, the floodplain, the uplands, up- and downstream, the shallow subsurface, and the deep subsurface. Water, sediment, organisms, nutrients, and contaminants move into, out of, and along the river channel via all of these pathways, but none of these exchanges occur equally through time and space. Biogeochemists describe hot spots where more chemical reactions occur than in other portions of a landscape, and hot moments during which these reactions are accelerated relative to intervening time periods. Analogously, connectivity along rivers is typically uneven. Much of the sediment carried downstream moves during floods. Organisms spawn, migrate, or hatch during specific seasons. Some portions of a river network have broad floodplains that store far more sediment and nutrients than do intervening portions of the network. Upwelling and downwelling between the channel and the subsurface occur primarily at obstacles such as beaver dams.

Along North St. Vrain Creek, about 75% of the total length of stream channels between the headwater lakes and the beaver meadow is composed of classic mountain streams: steep, whitewater cascades flowing swiftly between bedrock and large boulders that constrict the channel on either side, leaving little room for a floodplain to develop. Most of what enters these steep streams—water, sediment, wood, finer organic matter—keeps moving downstream. Here and there a logjam blocks the flow and creates a small backwater, or a particularly large boulder has a lee zone in which finer sediment and organic matter settle for a time. But mostly these segments of river are transport zones. Only about a quarter of the stream network is occupied by wider, more gently sloped valley segments, but these sections of river exert a disproportionate influence on the connectivity of the entire network. As stream flow enters these wider segments and moves more slowly, some of the materials being transported by the water drop from the water column to the ground. Big logs are

Six forms of river connectivity, with arrows indicating directions of movement of water, sediment, nutrients, and organisms.

more likely to remain in these valley segments, creating jams that trap organic material moving downstream. The twigs, leaves, and bits of bark and pine cone that form the organic matter are meat and drink to the microbes, bacteria, and aquatic insects in the stream. It's difficult to eat if the meat and drink is flying by and you're hanging on to keep from going downstream yourself, but the banquet is laid more temptingly in the slower-moving flow of the wide valley segments. Put more scientifically, even temporary storage for a few hours of organic matter in transit gives stream organisms a chance to ingest that organic matter, converting it into tissue that appears as greater abundance and diversity of organisms in the wider segments.

Logjams create wonderful storage zones, but the jams are not self-maintaining: if a key log breaks or is dislodged by high flow, the jam disappears. Beaver dams create even better storage zones, and beavers maintain the dams. My colleague Tim Covino and his graduate student Pam Wegener set up a grid of instruments that allowed them to measure stream flow in a steep, narrowly confined segment of channel upstream from the North St. Vrain beaver meadow and in the beaver meadow. The steep, narrow portion of stream efficiently conveyed water downstream throughout the snowmelt season. The beaver meadow, in contrast, stored water during the snowmelt peak, releasing the water gradually during the later summer and autumn, when the beaver meadow sustained higher base flows than the steep, narrow stream segment.

From the perspective of connectivity, beavers create downstream *dis*connectivity: their dams slow the downstream movement of water and anything being carried by the water. Beavers also create lateral and vertical connectivity. By forcing high flows over the stream banks, beaver dams increase the lateral connectivity between a channel and the adjacent floodplain. Enhanced overbank flows help to ensure a healthy river ecosystem. Floodwater infiltrates the ground and keeps the valley-bottom water table high. Sediment and organic matter deposited on the floodplain keep the water clear and clean in the channel and create germination sites for riparian plants and habitat for many types of animals.

The epigram "only connect" in E. M. Forster's novel *Howards End* refers to connections between individuals and social classes in 20th-century England. In natural environments, complete connectivity is not always a good thing. A river network composed only of steep, narrow valleys transmits floodwaters very quickly downstream, with little buffering or attenuation of the flood peak. This was exactly what happened during the September 2013 floods in the mountains around the North St. Vrain beaver meadow. As each tributary contributed its load to the rapidly increasing flood, the water ripped out roads and houses along the narrow valleys, eroding stream banks and causing the valley walls to unravel in landslides and debris flows. Slightly disconnected river segments, such as the beaver meadow, allowed the flood waters to spread across the valley bottom and lose some of their force in flowing through the dense thickets of willow stems and among the many smaller channels blocked by beaver dams. The same phenomenon occurs each year on a smaller scale during the spring snowmelt flood. As the floodwaters slow down passing through the beaver meadow, they deposit sediment and organic matter that contains carbon and nitrogen, thereby improving the water quality down-stream: beaver meadow as river kidney.

Any single beaver dam can be impressive, but it's typically a relatively small structure—not exactly Three Gorges Dam. But from many small dams come restored rivers, beaver meadows, and filled glacial valleys—the cumulative effects that change river networks.

May

Plugging the Nutrient Leaks

Just when spring appears to have arrived, a late-season storm blows down from the north. Despite the overcast sky, the temperature at first is beguilingly warm. Rain starts to fall, then changes to sleet as the temperature drops. "Rough winds do shake the darling buds of May," indeed. The sleet becomes graupel—crusty, rounded pellets of snow—and then wet flakes. Blobs of slush fall from overhanging branches and float briefly down the creek before melting and dispersing. Pulses of wind and snow gust in as birds shelter silently among the densely needled branches of big spruce trees. The moose that has spent the winter around the beaver meadow lies calmly in a protected spot at the base of a spruce. A foot of snow obliterates the newly green shoots of grass.

I see no outward indication of it, but perhaps, in the warm darkness of the lodge, the beaver kits have been born. They start small, only about a pound at birth, but they are born fully furred, with open eyes and incisors erupted, almost ready to get down to the business of chewing branches. Each year's litter is born in May or early June. Usually two to four kits are born, but a litter can be a single kit or as many as eight kits. Baby food for beavers is herbaceous vegetation, which the kits start eating within two weeks. By the end of July or early August the kits will be weaned and able to forage on their own. A varied diet of vegetation allows them to reach a weight of 10 to 16 pounds by the time the ice returns.

The first year is a grace period for the new kits. Unlike the yearlings, the kits do not help maintain the lodge or cache food against the lean days of winter. They simply get to enjoy life and explore the wondrous new world into which they have been born.

Diverse human observers watching this exploration by young beavers have interpreted their activities as an expression of joy. Hope Ryden thought more systematically than most scientists about the possibility that beavers experience joy and other emotions. Throughout her book *Lily Pond*, Ryden carefully distinguishes encoded or instinctive behavior from the ability to discover and apply new resources

Before the storm: filamentous algae (foreground, lower left) grow thickly in the warm, shallow, slowly moving waters of a side channel flowing toward the main channel, which is out of sight at the rear in this view. Down-valley is to the left, and the glacial lateral moraine is the forested ridge at the rear in this view. A color version of this figure is included in the insert section.

to a familiar situation. She writes of young beavers swimming about their pond, repeatedly diving and surfacing as perfectly coordinated with one another as synchronized swimmers. Ryden also describes an incident where she worried that the beavers at Lily Pond had grown too habituated to her presence. She decided to teach them fear of humans by smashing a shovel down onto the water surface the way a beaver slaps its tail to send an alarm signal. The adult beavers quickly responded negatively but the yearlings thought it was a game, repeatedly slapping their tails back at her until she became more aggressive and the adults shepherded the yearlings away.

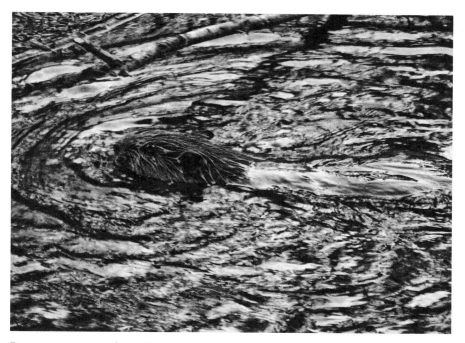

Beaver swimming in the Dall River, a tributary of the Yukon River in central Alaska.

The kits learn other important lessons during their first weeks out of the lodge, from how to repair a dam and construct scent mounds to how to groom themselves and each other. Kits clearly watch the adult beavers to learn how and what to eat. Ryden describes kits struggling with unwieldy lily pads, which sometimes ended up on a kit's head, until the youngsters watched the adults deftly roll the pad into a "cigar" that could be easily stuffed into a beaver's mouth.

Biogeochemical Engineers

The lily pad cigars provided the beavers of Lily Pond with vital nutrients. Nutrients are the components of food that an organism needs in order to survive. Two of the most important nutrients are nitrogen and phosphorus. Nitrogen is of particular interest to ecologists and biogeochemists for at least two reasons. On the one hand, nitrogen is critical for most living organisms, and limited nitrogen supplies restrict the growth of plants and ecological productivity. On the other hand, excess nitrogen supplies create a host of environmental problems and health risks for people.

The complex balance between necessary nitrate levels and excess nitrates reflects the many pathways that nitrogen follows in the environment. Nitrogen is continually exchanged among the atmosphere, soil, water, and living organisms. Atmospheric nitrogen gas (N_2) enters living organisms through the activities of either bacteria that live within the roots of legume plants or soil bacteria that convert nitrogen gas to ammonium (NH_4^+). Ammonium is converted to a nitrogen compound known as nitrite (NO_2^-) by other bacteria in the soil, but nitrite is unstable and quickly accepts another oxygen atom to become nitrate (NO_3^-). Nitrate is a nutrient that is readily available to plants.

By building dams that trap silt and clay and store water, beavers alter the amount of water and oxygen gas present in the soil and therefore the bacteria and plants living in and on the soil. Beaver dams also change the time required for dissolved and particulate forms of nitrogen to move downstream. Beaver dams thus alter the biogeochemical transformations of nitrogen. *Biogeochemical* sounds intimidating, but the compound word reflects the fact that chemical processes in natural environments are strongly influenced by geological processes such as erosion, sedimentation, and moisture, and by biological processes such as bacterial respiration. And behind all of these processes stand the beaver ecosystem engineers, in this context perhaps better described as biogeochemical engineers.

The first study to examine the effects of beaver ponds on the forms and quantities of nitrogen within a river system was conducted by Bob Naiman and Jerry Melillo in 1984. Working on the aptly named Beaver Creek in Quebec, Canada, they estimated a nitrogen budget for a riffle along the stream and for a beaver pond. A nitrogen budget is an accounting of gains, losses, and storage: how much nitrogen enters a section of stream from different sources, how much is stored within the stream, and how much leaves the stream section and continues down the valley.

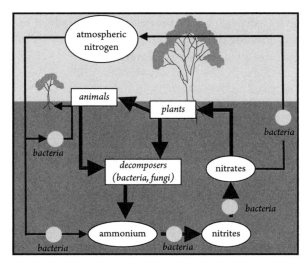

A schematic illustration of how nitrogen in different forms moves between the atmosphere, soil, plants, animals, and water. Each solid gray circle indicates a bacterially facilitated change in the chemical compounds containing nitrogen.

The nitrogen budget for a segment of stream strikingly illustrates how intimately the stream is linked to its surroundings. The budget starts with the inputs from the atmosphere, the surrounding vegetation, and the upstream river network. Some nitrogen enters a stream with rain and snow. This has always happened, although human activities that started with the Industrial Revolution have accelerated this "rain" of nitrogen in many regions. Any gardener knows that plants are not self-contained entities, but chemical reactors taking up and leaching compounds. Along a forested stream, rain and snow that fall onto the forest canopy can take up nitrogen from the tree leaves, so water that drips from the canopy into the stream can have higher nitrogen levels. Dissolved and particulate forms of nitrogen also enter a section of stream by flowing in from upstream portions of the river network. Plant litter such as leaves and twigs falling directly into the stream from the riparian forest or falling to the forest floor and then being carried into the stream contain nitrogen, as does large wood that enters the stream.

The second component of the stream nitrogen budget is the storage of nitrogen within water, sediment, and living and dead tissue. Within the stream, a tiny amount of nitrogen is stored in dissolved form within the water. Some of the nitrogen is taken up by aquatic plants and animals and stored within the tissue of these organisms. Dead wood can account for an important proportion of the nitrogen stored within a stream. Typically, a much greater quantity of nitrogen is attached to silt and clay particles and stored in bed sediment within the stream.

The final component of the budget is the outputs of nitrogen to the atmosphere or points downstream. Nitrogen leaves the stream segment by moving downstream in dissolved or particulate form or by release of nitrogen gas back into the atmosphere from the water. Nitrogen can also leave the stream within the tissues of aquatic insects emerging into the atmosphere, or by the chemical process known as denitrification. In denitrification, bacteria that live in environments with no oxygen gas facilitate the transformation of nitrate to nitrogen gas, which returns to the atmosphere.

When Naiman and Melillo compared equivalently sized areas of the stream in the riffle and the beaver pond, they found that the pond provided an environment in which microbes living in the bed sediment could effectively "fix" and store much greater quantities of nitrogen—about a thousand times more nitrogen than in the riffle. Subsequent research by Laura Lautz and her colleagues indicates that beaver dams also enhance the downwelling of surface water from the channel into the shallow subsurface, where other microbes can fix more nitrogen. Beaver dams and ponds create what biogeochemists refer to as hot spots—locations in which a disproportionately large amount of chemical transformations occur. This is extraordinarily important because of the problems caused by excess nitrogen in river systems.

Excess nitrogen comes from human activities, particularly runoff of fertilizer from crop fields, emissions from fossil fuel combustion, and runoff from feedlots

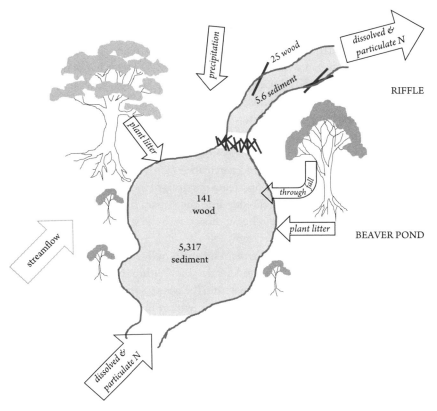

A schematic drawing of the results of Naiman and Melillo's nitrogen budget study on Beaver Creek in Quebec. The numbers indicate ounces of stored nitrogen per cubic foot of bed sediment or wood.

and concentrated animal feeding operations. Nitrogen from these resource uses, typically in the form of nitrates, enters rivers via surface water, groundwater, and both wet (with rain and snow) and dry deposition from the atmosphere. Rivers can remove at least some nitrates. Those all-important microbes that govern chemical transformations of nitrogen can break down nitrates within the soil beneath riverside plants, within the water just below the bed of the stream, and within sediments in wetlands. Wetlands—such as beaver ponds—are perhaps the single most effective type of nitrate cleanser because bacteria living in wetlands can remove as much as 80% of the nitrate load entering the wetlands. Not only do beaver ponds harbor these bacteria, but beaver dams force stream water down into the bed, allowing bacteria in the streambed sediments to remove further nitrates.

Where riparian zones and wetlands have been removed and excess nitrates enter rivers, the rivers become leaky with respect to nitrogen. Rather than being retained through burial in sediment and uptake by bacteria, the nitrogen is passed downstream. This can lead to at least three serious problems. The first is algal blooms. Algae happily take up nitrates, so extra nitrates can create a population explosion. As the algae die and decompose, processes of decay consume oxygen in the water, lowering dissolved oxygen concentrations below those necessary to sustain most animals. The resulting hypoxia—oxygen deficiency—can occur within slowly flowing portions of a river network but is most pronounced in nearshore areas off the mouth of large rivers. The Mississippi River is now a famous example of this, with a hypoxic or dead zone that, in some years, can cover an area of the Gulf of Mexico equivalent to the size of Connecticut.

The other two problems with excess nitrates directly involve human health. Excess nitrates in water can lead to blue baby syndrome, a condition more formally known as methemoglobinemia. This affects infants consuming water with high nitrate concentrations and can be lethal. In adults, high nitrate concentrations in drinking water can cause a form of cancer known as non-Hodgkin's lymphoma. Given these "side effects," the ability of beaver dams and ponds to remove nitrates from stream water and bury the nitrates in sediment looks both inexpensive and effective relative to human-built water treatment plants. Julia Lazar and several of her colleagues evaluated this issue by measuring the ability of beaver ponds in Rhode Island to remove excess nitrate from agricultural watersheds. Assuming a realistic density of 1.8 beaver ponds per square mile of watershed area, the scientists found that the ponds removed up to 45% of the nitrate in the stream. The net effect of beaver activities is thus to greatly increase the storage of nitrogen within valley bottoms.

Beaver meadows affect other elements besides nitrogen. Working on the Maryland coastal plain, David Correll and his colleagues found that beaver ponds reduced downstream levels of nitrogen, phosphorus, and carbon by up to 32%. Sequestering carbon in sediments underneath beaver ponds is also important in the context of climate warming because carbon dioxide is an important greenhouse gas. By flooding valley-bottom woodlands and creating oxygen-limited conditions in pond sediments, beavers can also initiate the formation of peat bogs and mires, which store large amounts of carbon. Colin Wells, David Hodgkinson, and Elizabeth Huckerby studied a prehistoric mire in northwest England that apparently formed as a result of beaver dams built about 2,800 years ago. Other studies demonstrate that the chemical transformations occurring within the bottom sediment of a beaver pond can raise the pH of river water and the water's ability to neutralize acids such as those brought in by acid rain.

Working with several other scientists, Bob Naiman studied the effects of beaver activity on Minnesota's Kabetogama Peninsula. During the period 1927–1988, beavers converted 13% of the 115-square-mile peninsula to beaver meadows and ponds. Prior to the beaver activity, chemical elements such as nitrogen, phosphorus, and carbon were held within the tissue of upland forest vegetation. When the beaver ponds inundated the forests, the trees died and these chemical elements were transferred to the pond sediments. The scientists found that few of the chemicals were carried downstream by river waters or returned to the atmosphere. Instead, saturation of the soil by the pond water created oxygen-limited conditions that fundamentally altered the biogeochemical pathways of the chemicals. When the beaver dams failed or the ponds filled with sediment and dried, the chemicals stored in the saturated soil became available as nutrients for new vegetation. During the 60-year period of beaver activity on the peninsula, beavers increased the levels of nutrients stored in the soil by anywhere from 20% to 295%, depending on the specific nutrient.

Oxygen-rich environments such as drier soils in which oxygen gas is present are chemically reactive. These environments, which dominate the uplands in most watersheds, can leak nitrogen and carbon to the soil water, groundwater, rivers, and the atmosphere. The thick, black, stinky muck of a beaver pond is a nitrogen and carbon bank, storing these vital nutrients in a form that can eventually be used by plants and animals when the pond dries or fills. This storage helps to prevent the nutrients from overloading the environment by passing rapidly downstream.

Carbon Stored Versus Carbon Lost

In a world in which accumulation of carbon in the atmosphere is causing global warming, the details of the global carbon cycle have become of great interest to many people. Here again, beaver engineering of river bottoms creates important changes in carbon movement and storage. In addition to measuring the effects of the North St. Vrain beaver meadow on stream flow, Pam Wegener and Tim Covino measured dissolved carbon and nitrogen in the confined channel segment of North St. Vrain Creek and in the beaver meadow. Their work provides insight into how beaver meadows progressively accumulate carbon and nitrogen. Like a doctor attaching electrodes to measure the electrical activity of a patient's heart for an electrocardiogram, Pam and Tim took the pulse of the North St. Vrain beaver meadow over the course of a year using arrays of instruments deployed across channels and ponds. They measured water level, water temperature, dissolved oxygen, and water chemistry at sites connected to the main channel by flowing water at the surface throughout the year; at sites that were connected only seasonally; and at sites with no surface connection.

Because flow paths in the beaver meadow are so complex and, in the case of subsurface flow, so slow, Pam and Tim could not simply measure stream flow into and out of the meadow. Instead, they mixed dissolved table salt into the creek upstream from the beaver meadow and used highly sensitive instruments to record the concentrations of salt at various downstream locations as a means of understanding rates and proportions of flow along different pathways. They needed sensitive instruments because they were not turning the beaver meadow into a salt marsh; they used only small amounts of salt that would not affect freshwater organisms.

Pam and Tim measured water temperature because flow through subsurface paths buffers daily fluctuations in water temperature. This can be used to distinguish water that has remained at the surface from water that has returned from below ground to the surface. They measured dissolved oxygen as an indicator of ecosystem metabolism. Changes through time in the oxygen dissolved in stream or pond water represent the balance among primary production, ecosystem respiration, and the exchange of oxygen with the atmosphere. Primary production is algal and other plant growth that releases oxygen into the water. Ecosystem respiration is the metabolism of organic matter by plants and animals, which removes dissolved oxygen. Water exchanges oxygen with the atmosphere in a direction depending on the oxygen saturation of the water. So, dissolved oxygen levels typically reflect the sum of two sources of addition (production and exchange) and one of subtraction (respiration). Finally, Pam and Tim used water samples to calculate daily fluxes of dissolved carbon and nitrogen into and out of the beaver meadow.

After a lot of field measurements, a lot of laboratory analyses, and a lot of mathematics, a health report emerged. The straight, narrow segment of North St. Vrain Creek upstream from the beaver meadow operates relatively simply. This portion of the creek is a watery highway, efficiently conveying water and dissolved organic carbon and nitrogen downstream, with minimal metabolic processing of dissolved nutrients. The beaver meadow is the HOV lane, retaining or releasing materials depending on the volume of flow. From May to October, when the greatest fluxes of water and nutrients are coming downstream with the snowmelt and rainfall runoff, the beaver meadow stores water and retains 99% of the dissolved organic carbon and all of the dissolved nitrogen entering from upstream. This phenomenal retention results from the diffusion of water and nutrients, via surface and subsurface pathways, across the beaver meadow. The area of surface water alone is eight times greater, per valley length, in the beaver meadow than in the upstream portion of the creek—and then there is all of the subsurface water storage. Increased surface area translates to increased removal of nitrogen by microbes and increased absorption and burial of carbon.

Metabolic processing in the beaver meadow slows during snowmelt peak flows but then picks up again as the flow recedes. Measurements in a side channel that was connected to the main channel by surface flows only during snowmelt peak revealed that primary productivity and ecosystem respiration were lower when snowmelt flows filled the side channel. However, productivity and respiration increased substantially once the side channel became disconnected from the main channel as snowmelt receded. This surge in metabolism was fueled by carbon and nitrogen delivered to the side channel during peak flow. By comparison, a pond that never had surface water connectivity with the main channel had consistently lower metabolic rates than the side channel.

In other words, the confined stream segment of North St. Vrain Creek maintained a steady flux of carbon and nitrogen downstream throughout the snowmelt season, whereas the beaver meadow stored the carbon and nitrogen. But this storage continued: levels of dissolved carbon and nitrogen downstream from the beaver meadow remained low throughout the year. We expected to find these patterns, but it was still exciting to see them appear so clearly. Natural systems are so complex and difficult to predict that it is common to put in a lot of hard work only to have equivocal results. Pam's thesis work was a delightful exception.

Beaver meadows, however, can also release disproportionate amounts of carbon to the atmosphere. In a 2014 scientific paper, Colin Whitfield and his co-authors estimated that there are now something like 30 million beavers worldwide. The majority of these animals (10 to 30 million) live in North America, but somewhere between 95,000 and 168,000 form an exotic population in Tierra del Fuego, and another million or so live in Eurasia. These beavers have created 3,700 to 16,000 square miles of ponded water. By making the boundary between ponded and flowing waters and adjacent vegetation more irregular and complex, beavers have increased the length of this ecologically rich zone by more than 124,000 miles.

The resurgence of beavers and the wetlands they create is reason for environmental celebration, but the Whitfield paper received a lot of attention for other numbers that it contained. The authors estimated that the resurgence of native beaver populations and their introduction into Tierra del Fuego accounts for methane emissions of 198,400 to 881,850 tons of CH_4 (methane) per year during 2000. This is about 200 times greater than methane emissions from ponds and flowing waters that became ponds circa 1900, when beaver populations in many regions reached a nadir. These numbers are highly imprecise because they are based on extrapolation from 10 site-specific studies, but the authors make an important point: relative to other wetlands, sediments in beaver meadows contain greater amounts of organic carbon per unit area and larger areas with microbes living in streambed sediments that can survive without oxygen. These microbes can create methane emissions through their respiration. This becomes important because methane is a powerful greenhouse gas.

The Whitfield paper, however, tells only one part of the story of carbon and beaver meadows. Measuring carbon storage in soils of beaver meadows in Voyageurs National Park on the border between the United States and Canada, Carol Johnston noted that much more carbon was stored per unit of area in the soils of the beaver meadows than in the soils of the adjacent upland forests. This carbon sequestration likely more than offsets the release of carbon to the atmosphere during the relatively short time that each beaver pond contains water.

My work and that of my graduate student DeAnna Laurel in beaver meadows within Rocky Mountain National Park support Johnston's findings. Our work focused on quantities of organic carbon stored within meadow sediments. We found that when meadows are abandoned by beavers and the meadow soils dry and erode, the organic carbon storage declines slightly, but it does persist through time at greater levels than in portions of the valley bottom never affected by beavers. Taken together, all of the major beaver meadows within the national park store an estimated 23% of the total carbon stored in soils within the park, even though the meadows occupy far less than 23% of the total surface area within the park.

Research thus far on the North St. Vrain beaver meadow indicates that snow-melt peak flows are the water- and nutrient-delivery period. The greater surface area and surface and subsurface connectivity of the beaver meadow relative to narrow segments of the creek provide abundant storage for water and nutrients. The nutrients are then processed by stream organisms once the stream flow levels re-cede. As part of this processing, some of the stored nitrogen and carbon is released downstream, but the volume released is much less than the volume that entered. Similarly, water stored in the beaver meadow is gradually released after peak flow, buffering fluctuations in downstream flow. The beaver meadow gathers the abun-dance of snowmelt, stores the riches in side channels and ponds and buried gravel, and then gradually releases water and some of the nutrients during the leaner times. A beaver meadow limits boom and bust scenarios by attenuating peak flow during floods, gradually releasing stored water during droughts, and maintaining higher baseline ecosystem metabolism and productivity. A beaver meadow: every creek needs one (or more).

June

The Thin Green Line

June, when the snows come hurrying from the hills and the bridges often go, in the words of Emily Dickinson. In the beaver meadow, the snows are indeed hurrying from the surrounding hills. Every one of the 32 square miles of terrain upslope from the beaver meadow received many inches of snow over the course of the winter. Some of the snow sublimated back into the atmosphere. Some melted and infiltrated into the soil and fractured bedrock, recharging the groundwater that moves slowly downslope and into the meadow. A lot of the snow sat on the slopes, compacted by the weight of overlying snow into a dense, water-rich mass that now melts rapidly and hurries down to the valley bottoms.

North St. Vrain Creek overflows into the beaver meadow, the water spilling over the banks and into the willow thickets in a rush. I can hear the roar of water in the main channel well before I can see it through the partially emerged leaves of the willows. Overhead is the cloudless sky of a summer morning. A bit of snow lingers at the top of the moraines. Grass nearly to my knees hides the treacherous footing of this quivering world that is terra non-firma. I am surrounded by the new growth of early summer, yet the rich scents of decay rise every time I sink into the muck. I walk with care, staggering occasionally, in this patchy, complex world that the beavers have created. I abruptly sink to mid-thigh in a muck-bottomed hole, releasing the scent of rotten eggs, but less than a yard away a small pocket of upland plants is establishing a roothold in a drier patch. A seedling spruce rises above ground junipers shedding yellow pollen dust and the meticulously sorted, tiny pebbles of a harvester ant mound.

I extract my leg with difficulty and continue walking. As I walk around the margin of another small pond, the water shakes. Sometimes the bottom is firm in these little ponds, sometimes it's mucky—I can't tell simply by looking through the water. Many of the small depressions have an oily water surface that looks fetid, but they are clearly good for frogs and bugs. Tan and brown striped wood frogs (*Rana sylvatica*) and chorus frogs (*Pseudacris triseriata*) sit quietly in the nearest

A caddisfly larva and case (resting on my palm) in the beaver meadow during June.

depressions. I stop to examine the bright orange algae that coats the submerged plant stems like a glove. A large insect larva, probably a dragonfly or damselfly, hangs suspended vertically in the water, its head downward.

Away from the noise of the main channel, I mainly hear birds. Chickadees, robins, warblers, tanagers, sparrows, and hummingbirds are nearby. A woodpecker knocks at a tree trunk up on the valley side slope. I have carelessly worn a red shirt, which occasions repeated close visits by hummingbirds that dart abruptly down to hover close to my face. I hear the whizzing flight of the hummingbirds before I see them, their movements too rapid to watch easily.

I climb up the berm surrounding an old beaver pond. These berms create ponds perched two to three feet above the drier surroundings. The gradually filling pond is rimmed by water sedge (*Carex aquatilis*), horsetails (*Equisetum* spp.), rushes, grasses, knee-high willows, and small mounds of sphagnum moss. The broad bowl of a mallard's nest rests on the ground back from the water's edge, the inside of the nest soft with downy feathers. The drake swims across the still water of the pond. In the shallows, I find a caddisfly larva case of flat sedge stems that reminds me of a Chinese pagoda. There are remarkably few mosquitoes, given the amount of standing water, and I watch with appreciation as dragonflies dart above the water. Numerous insects are hatching now, flying light as gossamer over the water striders dimpling the pond surface.

The down-lined nest of a mallard duck.

I move toward the main channel, crossing numerous smaller channels in which cold, swift water tinted pale orange-brown flows over clean sand. The fresh sand along the channel margins deposited by the September 2013 flood is already being hidden by grass, but piles of wood left by the flood remain. Some of the smaller channels flow at an elevation below the main channel; others are clearly perched above the main channel by at least a foot. The emerging leaves make it even more difficult to navigate across this complicated terrain in which water seems to be everywhere.

I do not see the beavers, but evidence of their activity surrounds me. Among all the other footprints and trails that crisscross the beaver meadow—deer, moose, elk, coyote, snowshoe hare—lie the low, broad troughs along which beavers have dragged pieces of wood. Commonly, these trails end at a freshly chewed stump and the route is lined with the piles of mud, grass, and broken branches that the beavers scent mark.

The beaver kits have now been born. Soon they will be playing in the world their parents have created. Enos Mills took play seriously. He considered it an important component of adult human life, and, in an era when scientists were more inclined to view animals as automatons or creatures completely driven by instinct, Mills paid attention to how juvenile and adult animals played. Scientists have subsequently

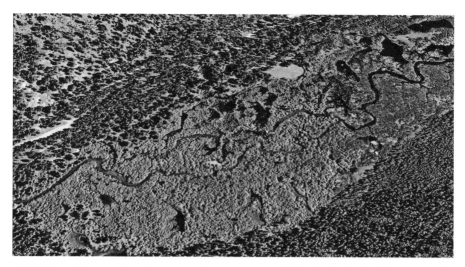

Aerial view of the meadow in summer. Image courtesy of Google Earth. A color version of this figure is included in the insert section.

caught up to Mills, recognizing the ubiquity and importance of juvenile and adult play among many animals.

The Beavers and the Forest

The adult and yearling beavers are also getting back to work after the relative inactivity of winter. There are dams and lodges to repair and caches of winter food to start. Other than humans, beavers are the only animal species in North America that can affect overstory vegetation by felling mature trees. There are few mature deciduous trees to be felled in the North St. Vrain beaver meadow except for aspens along the margins of the meadow, but plenty of thick willow stems can be harvested. This might seem to be a bad situation for the willows, but their abundance in the beaver meadow suggests that they cope quite well with beaver harvest.

In the give and take between beavers and riparian trees, beavers take existing trees but provide germination sites for seedlings. Studying sedimentation behind beaver dams in eastern Oregon, Michael Pollock and other scientists documented the changes that occurred when beavers dammed a deeply cut channel. The scientists found that sediment accumulated behind dams at rates as fast as 1.5 feet per year during the first six years after a dam was built. As the channel gradually filled, rates of sediment accumulation dropped to less than a third of a foot per year. Sediment also started to accumulate on surfaces beside the channel. With time, the backwater associated with the beaver dam resulted in a broadly spreading wedge

Beaver trail (above) and freshly felled tree. A color version of this figure is included in the insert section.

One of the hidden passages of water in which the North St. Vrain beaver meadow is rich.

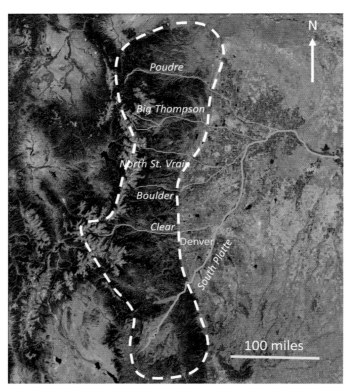

Location map of the Colorado Front Range (outlined in white) and the major tributaries of the South Platte River. The South Platte flows eastward to join the Platte River in Nebraska.

A portion of the beaver meadow in early June, looking upstream toward the Continental Divide.

The same pond, along the southern margin of the beaver meadow, covered with ice in November.

Looking upstream along a side channel of North St. Vrain Creek. The willows along the valley bottom give way to conifers on the valley side slopes.

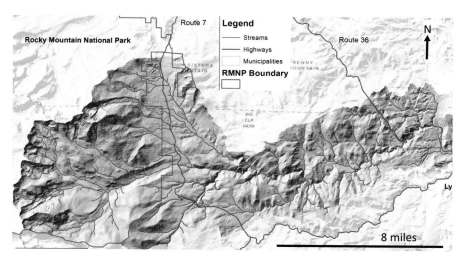

Shaded relief map of the North St. Vrain Creek watershed. The boundary between Rocky Mountain National Park and Roosevelt National Forest bisects the basin, and the town of Lyons is at the base of the watershed, along the mountain front. The western boundary of the watershed is the Continental Divide.

View of the beaver meadow from the Sandbeach Lake trail. Lateral moraine at rear (down valley to the left).

Upstream view of a beaver dam on a side channel in the North St. Vrain beaver meadow.

A beaver pond and lodge in Colorado, with inset view showing the beaver canal dug into the bed of the pond as a darker band across the center of the pond.

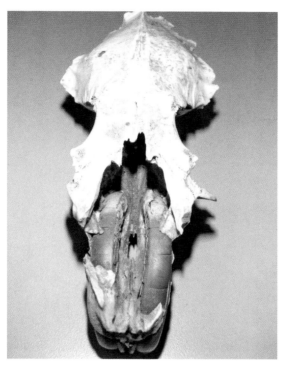

Top of a beaver skull. The incisors form a semicircle from the base of the eye sockets to the protruding tips at the front of the animal's mouth.

An abandoned beaver dam that North St. Vrain Creek, at right, has cut through. The dam is a mixture of beaver-gnawed branches, finer organic matter, and sediment.

Snow mounded along a side channel of North St. Vrain Creek. The glacial lateral moraine appears as a forested ridge in the background.

A cutbank about four feet tall along Fish Creek after the September 2013 flood. This site, upstream from the dam that beavers built along the creek in 2011, has the thick, dark, organic-rich soils characteristic of a beaver meadow, although at present the site is a dry pasture.

One of the main beaver ponds, with a lodge in the center. The ice and snow are gone and the pond is open, but the willows have not yet begun to leaf out.

Before the storm: filamentous algae (foreground, lower left) grow thickly in the warm, shallow, slowly moving waters of a side channel flowing toward the main channel, which is out of sight at the rear in this view. Down-valley is to the left, and the glacial lateral moraine is the forested ridge at the rear in this view.

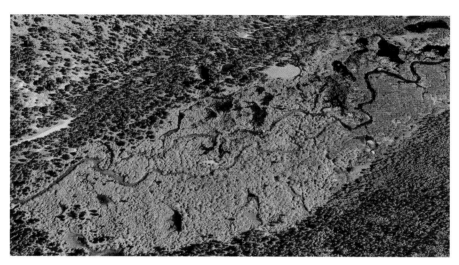

Aerial view of the meadow in summer. Image courtesy of Google Earth.

Beaver trail (above) and freshly felled tree.

Water sedge (above) and willow buds in the beaver meadow during June.

The still waters of the beaver pond mirror the adjacent trees.

A wood frog in the North St. Vrain beaver meadow.

A portion of the North St. Vrain beaver meadow in August: not quite parching, as wetland plants encroach on the steadily retreating standing water.

A beaver harvesting willow branches along the Dall River in central Alaska.

Beyond the edge of the drier terrace, the beaver meadow remains wet and green in August.

Elk grazing in Moraine Park during 2011.

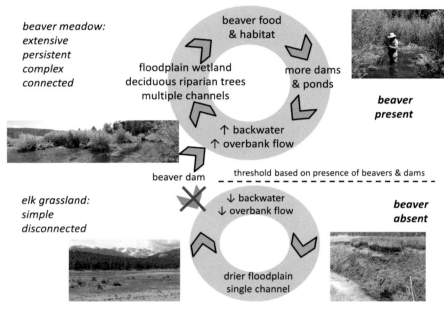

An illustration of beaver meadows versus elk grasslands and the feedbacks that occur to maintain each of these conditions.

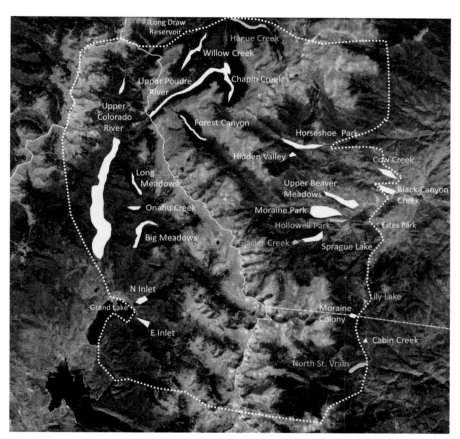

Schematic map of sites historically (yellow) and currently (green) occupied by beavers in the vicinity of Rocky Mountain National Park (approximate park boundaries in dotted white line).

The beaver meadow in mid-October, with snow on the high peaks and the willows and grasses in their autumn colors.

Bull moose at the beaver pond.

Before the September 2013 flood: a for-sale sign in front of the beaver pond and lodge along Fish Creek.

Soil displaced by pocket gophers lies revealed on a patch of bare ground beneath the pine groves bordering the beaver meadow. The pencil provides a sense of scale.

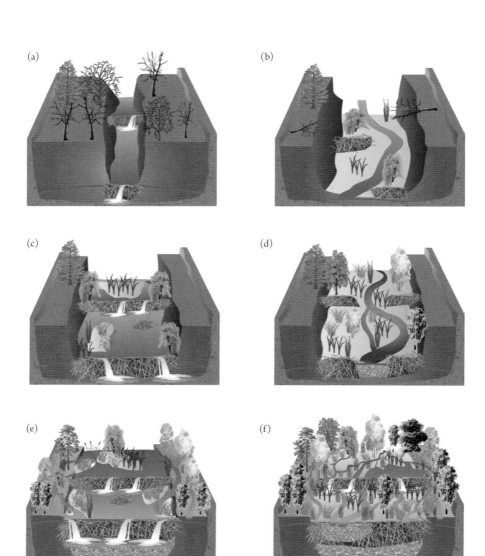

Schematic illustration of how beaver dams alter streams that have cut down into valley sediments. (a) Beaver dams in narrow streams are more likely to fail during high stream flows. (b) This results in erosion around the dam helps to widen the incised valley, allowing a narrow floodplain to form. (c) The widened valley results in less flow energy during floods, which allows beavers to build wider, more stable dams. (d) Streams that are incised commonly transport substantial sediment. Beaver ponds fill rapidly with sediment and the beavers move to another site along the channel, but the pond fill provides germination sites for vegetation. (e) This process repeats itself until the beaver dams raise the water table enough to reconnect the stream to its former floodplain. (f) Eventually, vegetation and sediment fill the ponds and the stream ecosystem becomes complex and resilient. (From Pollock et al., 2014, Figure 4.)

The history of water movement as recorded in ice, December in the beaver meadow.

Winter seals the beaver pond.

of newly deposited sediment upstream. Eventually, the valley bottom upstream from the beaver dam had five times more area within two feet in elevation above the channel. This newly created land was prime habitat for riparian plants.

Riparian woody plants such as willows and alders are among the pioneers that germinate first in newly deposited river sediment. These early successional species need high levels of nutrients in the soil, which is exactly what is provided by newly deposited river sediment mixed with nitrogen-rich plant litter carried by the river. In the absence of beavers, the shade provided by willows can allow the seedlings of more shade-tolerant woody plant species to germinate and eventually take over a site. These shade-tolerant, late successional species, which are mostly conifers, are able to survive low soil nutrient levels that are maintained by the slowly decomposing litter dropped by the conifers. The continuous harvesting of early and mid-successional species by beavers can reverse the progress of forest succession. By increasing light penetration, beaver foraging creates an environment suitable for early successional willows and alders. In a sense, the beavers keep the riparian forest ever young. In portions of the valley bottom farther from the water's edge, beavers can hasten forest succession by limiting understory competition for pine and spruce seedlings.

The riparian woody plants that beavers prefer to eat have their own adaptations to beaver harvest. Narrowleaf cottonwood (*Populus angustifolia*) and aspen (*P. tremuloides*) that are heavily harvested by beaver start to reproduce by producing adventitious buds at the roots. These buds sprout around the base of the cut plants. Beavers do not eat these newly sprouting plants because the sprouts have elevated concentrations of a chemical compound that presumably does not taste very good to beavers. This helps to explain how beavers and their preferred food plants maintain some balance in riparian forests.

In contrast, willows browsed by beavers during winter simply grow more stems and branches, and this new growth is rich in nutrients—added value for the beavers. The study that demonstrated this phenomenon, undertaken by Annelies Veraart and her colleagues at a national park in the Netherlands, would presumably go down in beaver history if the animals kept annals. The scientists built "cafeteria racks" of bamboo poles positioned close to the water's edge and attached bundles of willow shoots to the poles. I assume the beavers appreciated the effort, but it's not reported whether the animals became lazy and complacent.

The Thin Green Line

The 20th century was not a good era in which to be a wetland plant. People have been draining wetlands since the start of agriculture. Much of this drainage occurred along river bottomlands that were seasonally flooded, partly as a result of beaver activities and partly as a result of seasonal high flows that simply overflowed the river channel even in the absence of obstructions such as beaver dams. George Washington's library included a copy of *Practical Treatise on Draining Bogs and Swampy Ground*, first printed in 1775. Scientists estimate that the conterminous United States included approximately 221 million acres of wetlands—an area more than twice the size of California—at the time of European settlement. The first settlers got down to work almost immediately cutting forests and draining marshes and swamps. The process of systematically draining wetlands accelerated dramatically, however, during the 20th century. More than 50 million acres were drained for croplands in the Midwestern United States alone during this century. By the mid-1980s, six states (Ohio, Indiana, Illinois, Iowa, Missouri, and California) had lost more than 85% of their wetlands. The states within the upper Mississippi River drainage network were largely drained during the period after 1800, whereas California's wetland loss occurred mostly after 1860. Another 17 states, including Colorado, had lost 50% to 85% of their wetlands by the mid-1980s. The situation is similar in Canada, where 65% to 80% of coastal marshes have been destroyed and extensive portions of Ontario and the prairies have lost more than 70% of their wetlands.

Wetlands have been drained primarily for agriculture. The high water levels that allow organic matter to accumulate and slowly decay create exceptionally fertile soils and, along rivers, seasonal floods replenish the soil organic matter. Wetlands have also been drained because they are perceived to be in the way of construction of roads, housing developments, and other municipal and industrial infrastructure. And wetlands have been drained and paved because, except for environmental scientists and birdwatchers, wetlands have few fans. Think of all the negative connotations of wetlands—mosquitoes, flat, boring, difficult to move around in, and unhealthy (the name for the disease malaria reflects the poisoned air thought to emanate from wetlands).

Public perceptions of wetlands are diametrically opposed to those of environmental scientists, who view wetlands as some of the most diverse and productive ecosystems on the planet. What might appear as a monotonous expanse of shallow water resolves at closer view into a rich array of habitats. Subtle differences in elevation and soil moisture translate to differences in the microbial communities inhabiting the soil, the biochemical reactions occurring within the soil, the plants that can germinate and flourish, and the insects, birds, and other animals that live among or feed on those plants.

Wetlands are also in some sense the kidneys of the landscape, particularly wetlands along river valleys. Water flowing into the wet valley bottom from the adjacent uplands, or overflowing from the river channel during floods, carries sand, silt, and clay, as well as dissolved chemicals. As the water spreads across the wetlands, the velocity of flow declines, allowing some of the sediment in suspension to settle onto the surface: this is how rivers build floodplains. Attached to the sand, silt, and clay are nutrients such as nitrogen and phosphorus. Wetland plants can take up these nutrients through their roots and pass the nutrients on to the herbivores that eat the plants. Some of the surface water filters into the soil and moves more slowly in the subsurface, allowing microbes and bacteria to take up some of the dissolved chemicals in the water, including dissolved carbon and nitrogen.

A variety of wetlands can occur in diverse settings. Salt marshes and other coastal wetlands are common along low-lying coastal regions. Inland marshes form in depressions where deep standing water occurs at least seasonally and rooted vegetation emerges from the standing water. Sedge meadows form where the soil is saturated but not submerged beneath the water surface and wet meadows form where the soil is seasonally saturated. Depressions created by continental ice sheets now hold standing water on the northern plains of the continental United States, forming prairie potholes. In mountainous settings, fens occur at sites such as the toes of mountain slopes and alluvial fans, where groundwater reaches the surface year-round, but little sediment is eroded or deposited. Fens are a type of mire, as are bogs. Water enters bogs mostly from rainfall. Dead plant material—especially sphagnum moss—accumulates in bogs to form peat. Salt flats develop in depressions that receive either surface runoff or upwelling groundwater but also dry out each year, allowing the water to evaporate while salts are left behind. Swamps or bottomland forests develop where trees grow in saturated soils. The largest single category of wetlands, however, are those forming along river valleys.

After two centuries of intensive and widespread alteration of river valleys, I find it difficult to even imagine how different the United States looked in 1650 AD, but I know that there was much, much more standing water on the surface. Consider a few facts and figures. Every historical description by the first people of European descent to reach a river flowing through forested lands in North America—from New England to the Deep South, the upper Midwest, and the Pacific Northwest—emphasizes how hellish it was to move downstream because of the enormous volumes of wood in the channel. Individual snags partly sunken in the streambed and pointing upstream; sawyers sunken in the streambed and pointing downstream; sweepers leaning out from the banks to sweep unwary boatmen from their vessel; logs floating downstream with the force of battering rams; logjams extending all the way across the channel; and even enormous accumulations of wood such as the 40-mile-long Great Raft on Louisiana's Red River, estimated to have been lodged along the river for 400 years. All of these obstructed the movement of water and boats downstream. Individual trees toppling into rivers and masses of trees brought in by landslides, wildfires, and hurricanes all concentrated along the river bottoms. The wood made navigation difficult but also created floodplain wetlands. Just like a beaver dam, individual logs and accumulations of wood helped to pond water upstream, as well as forcing high flows over the channel banks and out onto the floodplain, where the water moved more slowly downstream, infiltrating into the ground and collecting in ponds and lakes. In flat country, these effects could extend for tens of miles, as reflected in Eliza Steele's description of crossing the wet Illinois prairie during the late 1830s.

More than half of the area once covered by floodplains and wetlands in the lower 48 states was drained by the year 2000. Wetlands of the lower Mississippi River and Gulf states covered about 78 million acres—35% of the total wetlands in the lower 48. These wetlands were 120 miles wide at the Mississippi floodplain's widest point and extended as far north as the bottomland forests of the Cache River in Illinois. Similar bottomland forests extended up to 200 miles inland along river valleys all around the Gulf of Mexico from the Neches River of Texas east to Florida's Caloosahatchee River.

The U.S. Army was initially charged by Congress with maintenance and improvement of inland waterways for navigation and flood control. Records kept by the Secretary of War indicate that more than 1.8 million logs were removed from major navigable rivers in the conterminous United States between 1870 and 1912. At the same time, rivers were straightened, dredged, and lined with levees—activities collectively known as channelization—to keep floodwaters within a single channel. The combined effect of these activities was to cut off the water supply to floodplain wetlands. Floodplains were also actively drained by cutting ditches through them and installing tile drains to lower the water table.

The U.S. Army worked on the big rivers, and businesses and private landowners worked on the smaller rivers. Loggers were particularly effective in clearing wood from streams and rivers in order to use the river network to float cut timber downstream to collection booms. Similarly, farmers were particularly effective in dredging even the smallest creeks that could flood adjacent bottomlands. For example, in the Des Plaines, Kankakee, and Sangamon River basins, all tributaries of the Illinois River, every tiny creek was channelized. And, of course, fur trappers initially, and subsequently farmers and others trying to limit overbank flooding, were busy removing those estimated 400 million beavers that lived and built dams throughout the small to medium rivers within each drainage basin. An estimated 51.1 million acres of beaver-flooded land in Illinois had been reduced to 511,290 acres by 1990.

Channelization, beaver trapping, and wetland drainage created the desired effect of improving navigation, limiting overbank flooding, and opening millions of acres of fertile bottomlands to agriculture and human settlement. Channelization and wetland drainage also created a plethora of undesirable effects. Nutrients coming off the uplands, especially excess nitrogen fertilizer from crop fields and feedlots, were no longer filtered and retained by the wetlands. The newly efficient channels sent large volumes of water and sediment quickly downstream, exacerbating flooding where the channelization ended. Instead of moving slowly across the floodplain, filtering into the groundwater, and maintaining high water tables, heavy precipitation and floodwaters were shed off landscapes, leading to progressive drying and consequently greater vulnerability to droughts and fires. As floodplain and wetland habitat disappeared, the numbers of fish, waterfowl, and other animals declined. In Illinois, early settler Albert Herre wrote of swarms of waterfowl filling the air, until "the advent of tile drainage in the early [18]80's completed the transformation of the prairie into ordinary farm land . . . Of course the ducks and geese stopped coming, for there was neither water nor food to attract them . . . The crawfish and bullfrogs disappeared in a hurry . . . the prairie as such disappeared, and of course its characteristic life with it" (p. 80).

The cumulative effects of channelization and wetland drainage conducted across an entire watershed are exemplified by the Mississippi River, which was once the nation's premier floodplain river. From the prairie pothole wetlands of Wisconsin, the Dakotas, and Iowa, through the broad floodplains of the tributary Ohio and Illinois Rivers, and down the extensive lowlands of the central and lower Mississippi River itself, the river network historically included an enormous extent of wetlands. By 1978, less than 20% remained of the floodplain forest that historically covered 40,100 square miles in the lower Mississippi River valley from Cairo, Illinois, down into Louisiana. Even less than 20% of the floodplain wetlands remain. Like a person who has lost most of both kidneys, the river no longer functions well. Water, sediment, nutrients, and contaminants entering the river through a million tributaries large and small are not stored in and filtered through floodplain forests, meadows, and wetlands, but are instead flushed quickly and efficiently downstream and into the Gulf of Mexico. In the Gulf, the contaminants, particularly excess nitrogen shed from over-fertilized crop fields, create a dead zone that can extend across 6,500 square miles and in which few living things can survive. This situation is not unique to the Mississippi River. Most estuaries and nearshore zones associated with major rivers in high-income countries have similar problems: the Chesapeake Bay, the lower St. Lawrence River estuary, and parts of the Baltic Sea and the Black Sea are other notable examples.

Increasing concern since the 1970s over these unintended consequences has led to efforts to protect and restore wetlands and floodplains. Sometimes this is as simple as setting aside still-functioning wetlands in a national wildlife refuge or a private preserve administered by a nonprofit organization such as the Nature Conservancy. Setting aside land is never really simple, of course, except in comparison to the complexities of restoring wetland function where the wetlands have been drained and isolated from the river by levees and the river itself no longer has large peak flows because of dams and diversions.

Some of the natural processes that create wetlands take a very long time to restore. Old-growth forest is particularly effective at creating large, closely spaced logjams that span streams and enhance overbank flooding, but a forest typically requires 200 years to attain old growth characteristics. Beavers are faster.

Green Engineering

Studies from diverse beaver meadows demonstrate how beaver ponds create good habitat for aquatic plants. Mixed in with the mineral sediments on the bed of the pond is plant litter rich in nutrients such as nitrates, silicate, and phosphate. Alan Law and his colleagues described the changes in Scottish headwater streams draining pasturelands after a pair of beavers was reintroduced to the site in 2002. Within just over a decade, the beavers built 10 dams along a mile of stream and the retention of organic matter increased 7-fold, while the biomass of aquatic plants increased 20-fold.

A study of beaver ponds in Minnesota by A. M. Ray and others demonstrated that ponds are first colonized by free-floating, easily dispersible plants such as duck-weed (*Lemna* spp., *Spirodela* spp.), water-meal (*Wolffia* spp.), and bladderwort (*Utricularia* spp.). These plants are most abundant in ponds four to six years old but become less abundant as a pond ages. By the end of the first decade in the exist-ence of a pond, rooted aquatic plants such as pondweeds (*Potamogeton* spp.), horn-wort (*Ceratophyllum demersum*), and waterstarwort (*Callitriche vulgaris*) colonize the pond, becoming most abundant in middle-aged ponds (10 to 40 years old) at the Minnesota site. As the pond grows older, it gradually fills with sediment and organic matter and becomes less connected to surface flow from adjacent channels. Floating-leaved, rooted aquatic plants such as watershield (*Brasenia schreberi*), water lily (*Nymphaea* spp.), and yellow water lily (*Nuphar lutea*) appear last. These plants are not so readily dispersed between suitable sites as some of the early plant colonists of the pond, but, once introduced, the floating-leaved plants increase in abundance until the pond is about 40 years old.

Ray and colleagues found that the number of plant species in a pond increases steadily during the first 40 to 50 years, then levels off or declines. The lack of fur-ther increase in plant diversity probably reflects competition for light between the floating-leaved plants, which form a canopy on the water surface, and submerged vegetation. The leveling off in number of species may also reflect the effects of all the animals that eat aquatic plants, including waterfowl and beaver. When the pond finally dries out, increased oxygen levels in the pond sediment allow chemical reactions that release the nitrogen and phosphorus stored in the sediment. Soils develop on the bed of the pond and grasses, sedges, and shrubs colonize the site.

A valley bottom inhabited by beavers is a dynamic environment. Dams are built, ponds form and then gradually fill, and the beavers build another dam elsewhere. Trees are harvested in one site, then the beavers shift their harvesting elsewhere. At any point in time, a wide variety of germination sites are available for plants to colo-nize. The diversity of soil moisture, soil nutrients, and sunlight reaching the ground surface explains the diversity of plant species measured by scientists. In one study in New York, beaver activity increased the number of species of herbaceous plants by over a third relative to valley-bottom sites without beavers. In North Carolina, wetland habitats created by beavers supported plant species not found elsewhere in riparian zones and maintained host plants needed by a rare species of butterfly. In the Adirondack Mountains of New York, beaver activities so enhanced the di-versity of plants present in the valley bottoms that, when scientists tallied plant

Water sedge (above) and willow buds in the beaver meadow during June. A color version of this figure is included in the insert section.

species across the entire landscape, the beaver activities had increased the number of plant species by 45%. Beaver ponds occupied by aquatic plants in the Adirondack study area eventually developed into meadows dominated by the grass species *Calamagrostis canadensis*, as well as sedges (*Carex* spp.) and stands of speckled alder (*Alnus incana*).

"Green" is now commonly used as a descriptor for more environmentally focused versions of some activity. Hence we now have green political parties and voters (sometimes known simply as "greens"), green energy, and green engineering, among other things. Beavers are the original green engineers, building environmentally beneficial structures that create a wide array of secondary effects, including literally enhancing the greenery of river corridors.

The green line that represents native riverside vegetation has thinned to the point of snapping along many rivers in the United States and around the world. As the green line has vanished, so have all of the processes and organisms dependent on its presence, from slowing the downstream passage of floodwaters and retaining fine sediment and nutrients, to providing habitat and food for microbes, plants, insects, amphibians, fish, reptiles, birds, and mammals. The parallel blue and green lines of rivers and riverside vegetation stitch together landscapes. We cannot weaken or rip out those threads without seeing larger swathes of landscape come apart as their ecological functions diminish. Beavers are not the answer to all problems, but their enrichment of river corridors—their ability to widen and strengthen the green lines—is a very good start.

July

Of Fish and Frogs and Flying Things

By mid-July, abundant water continues to move in all directions within the beaver meadow. Water flows noisily down the main channel, creating deep pools where it mixes with water entering from secondary channels. Deeper waters well up from beneath overhung banks and the willow stems along the banks remain partly submerged. Pieces of driftwood collect where the channel bends, floating in perpetual circles atop the shadowed water. The water is clear of suspended sediment but stained slightly brown. Flow is noticeably lower in the secondary channels, where algae and bacteria stain the cobbles reddish-brown. Shallow water runs down a beaver trail toward the main channel, and I can easily imagine the trail eroding into a small canal over a period of years.

The fern-like stems of rust red that grew beneath the pond waters earlier in the season have now emerged and bloomed, revealing a row of pink flowers of elephant's head. Diminutive white twinflowers bloom near the conifers at the edge of the meadow. Stalks of pink and white Pyrola flowers rise above their ground-hugging leaves, which have been green since April. Mountain bluebells form clusters of indigo among the green hues of the grasses and sedges. Broad white blossoms of cow parsnip create a canopy above the other herbaceous plants. Aptly named shooting stars resemble tiny bursts of yellow and white trailing spiraling pink petals as they lean over the ground.

The songbirds are less vocal than in June now that they are busy tending to nestlings weak at flying, but I can still hear the notes of chickadees, sparrows, and warblers, underlain by the distant croaks of ravens. Hummingbirds continue their mating displays, diving toward the ground as though intent on suicide, only to pull up at the last moment. The red blazes on their throats flash like fragments of momentary flame amidst the thick greenery. Mosquitoes are more noticeable now, despite the damselflies and dragonflies busily hunting back and forth across the openings among the willows. Beds of matted grass lie dispersed across the meadow. I watch two leggy young moose easily cross the swift flow of the main channel. I am

An underwater view of the edge of the main channel, looking downstream. The grasses and willows growing along the steep bank, at left, are partly submerged, creating hiding spots for insects and fish. Air bubbles form lighter spots within the water and just under the water surface at the top of this view.

not overwhelmed by the sense of exuberant, abundant life, as I was in June, but the beaver meadow remains a busy place.

The beavers have repaired the breached ponds of winter, and I can see my winter footprints in the bed of one pond now brim-full. Beaver trails densely marked with scent mounds radiate in all directions from the pond. I follow a trail, crossing the narrower paths made by moose and deer, to a stand of aspen. Elk and beavers have been at work here. Swaths of tooth-scraped bark a few feet above the ground reveal the winter browsing of elk. Beavers have left pale, sharpened stumps and peeled branches littering the ground.

I continue on to the main pond, which remains full to overflowing. Cut and peeled branches glow white beneath the quiet, tea-colored water at the edge of the pond, where frogs cease calling and leap abruptly for cover as I approach. Water striders large and small skate across the water, the smaller ones barely dimpling the surface. I sit down to watch, slapping repeatedly at the mosquitoes and tiny flies that sit down with me. The air is busy with insects, not all of which bite. A swallow-tail butterfly moves among the pink fireweed and purple asters blooming beside the water.

Juvenile moose in the shallows at the edge of the main channel.

By the end of this month, the beaver kits will be weaned and able to forage on their own. This intricate world of ponds, meadows, and willow thickets that their ancestors have created must be a place of wonder for the kits. The scent mounds maintained by their family mark the boundaries of the known world, but inside those boundaries lies a smorgasbord of salad—dozens of varieties of forbs and grasses, each of which the kits now taste for the first time as they steadily grow from their birthweight of a mere pound. The adult and yearling beavers stay busy, putting on their own pounds against the lean period of winter, caching peeled stems and branches in the pond waters, maintaining the dams and the scent mounds. The yearlings also help the adults in feeding, grooming, and guarding the kits. What a nice life for the kits—a new world to explore and a whole family to care for them. I hear the slap of a tail as I move slowly around the edge of the pond, but all I see are the spreading circles where the beaver disappeared beneath the water.

The beaver is likely to resurface quietly at some place where I will not notice it: the pond is shallow and relatively small. The physiology that allows the beaver to dive is impressive. By forcibly immersing beavers in water, scientists have measured the changes that occur in the animal's body. Diving forces a beaver to conserve limited oxygen. The animal's heart rate slows, which saves oxygen. Blood rushes to the brain and heart, increasing the flow of oxygen-rich blood to the brain. A beaver can tolerate high concentrations of carbon dioxide in its tissues. Beavers can also

Beaver-gnawed stump (top) and elk-scraped bark (bottom) in an aspen grove along the margins of the beaver meadow.

truly deep-breathe: a beaver can exchange as much as 75% of the air in its lungs, whereas a human can exchange only about 15% during each cycle of exhalation and inhalation.

Diving also requires special adaptations for maintaining body temperature. Beaver fur is compressed in water, so insulating air between the hairs escapes. Fundamentally, a beaver remains a land animal that is not well adapted to a prolonged stay in water, unlike an animal such as a seal. A beaver's body fat provides better insulation than its fur. However, increased blood supply to the working muscles of a swimming beaver expands the thermal core of its body, causing higher rates of convective heat loss to the surrounding water. A beaver's tail and hind legs are particularly important as heat exchangers. The base of the tail contains a network of blood vessels that regulate this heat exchange. Within the tail, arteries carrying warm blood from the core of the animal's body transfer heat to adjacent veins and the veins carry the heat back to the core. Blood at the surface of the tail is cooler, so that little heat is lost to the surrounding water. During summer, a beaver can lose a quarter of its body heat via its tail, but during winter the patterns of blood flow alter so that as little as 2% of the animal's body heat is lost.

As expected, I do not see where the beaver resurfaces. The pond is quieter than the channels. Beyond the calls of the songbirds I hear distant thunder from a storm moving swiftly down the valley. Thunderheads can appear magically out of a clear blue sky at this time of year, reaching a crescendo of towering, muscular white columns by noon. Now raindrops start to fall from the humid air, gently dimpling the pond surface at first, then coming hard and fast. Each drop splashes back upward, silver against the dark water. When the rain stops, I realize how loud the splashes were. As the water surface once more becomes placid, I see the circles of feeding fish. The beaver meadow is prime fish habitat and at least a few of the numerous human visitors now streaming into the park eddy out into this quiet little maze of channels and ponds to try their luck at angling.

I think of Lewis Carroll's lines from *The Walrus and the Carpenter*: " 'The time has come', the Walrus said, 'to talk of many things: Of shoes—and ships—and sealing wax—of cabbages—and kings . . .' " For the beaver meadow, the time has come to talk of fish and frogs and flying things.

A Fish in Every Pond

Once upon a time, fisheries managers advocated removing logjams and beaver dams to facilitate the movement of fish along a river. The reasoning seemed to be that jams or dams completely spanning the channel formed an impassable barrier that fish swimming upstream could not jump over or swim around. I wonder how the fisheries managers thought billions of inland fish thrived for millennia in the company of numerous logjams and abundant beavers. Subsequent research demonstrated that many species of fish regularly cross beaver dams in both up-stream and downstream directions. Beaver dams only occasionally create barriers to fish movement during low flows.

Once upon a time was actually not very long ago. Although some fisheries managers still remove beaver dams, most stopped doing so within the past 20 years. Today many fishery biologists advocate exactly the opposite: leave the logjams and beaver dams in place and, if possible, add more.

The turnaround is not particularly difficult to understand, although it was challenging to collect the data to support the change in attitude. Studies of logjams and individual pieces of wood in streams began in the Pacific Northwest during the late 1970s, spurred in part by the need to understand stream ecology in a region where extensive timber harvest and other activities seemed to be driving once-plentiful salmon to extinction. By the end of the 20th century, numerous studies in the Pacific Northwest and around the world clearly demonstrated that, far from impeding fish movement, the presence of wood in streams created vital fish habitat. Biologists census fish populations using electroshocking. The biologists go along a stream with buckets, shocking a portion of the channel, catching the stunned fish in the buckets to count and measure them, and then returning the fish to the stream. As my colleague Kurt Fausch described it, all goes well until the biologists reach a portion of the stream with abundant wood, at which point they don't have enough buckets to hold all of the fish that are present.

Wood in streams, whether individual pieces or jams, at least partly blocks the downstream flow of water. Some water is stalled upstream from the wood, forming a backwater pool in which finer sediment and organic matter settles to the streambed. The backwater provides good resting or feeding habitat for fish. As the flow accelerates going around or over the wood, it scours the streambed and bank, creating another pool downstream or overhung banks that fish also like. Streams with abundant wood typically have larger and more numerous pools than otherwise similar streams without wood. These pools can provide refuges for fish when shallower portions of the stream freeze in winter, grow too warm in summer, or simply cease to flow during dry periods. Wood also creates overhead cover for fish, helping to shelter them from predators.

Wood in streams also provides a fine home for algae, bacteria, microbes, aquatic insects, and sometimes even rooted aquatic plants. Some species of fish like to eat the aquatic plants or insects, and at least one group of fish—the armored catfish of South America (*Panaque* spp.)—eat the decaying wood, although it remains unclear how much nourishment they get directly from the wood and how much from algae growing on the wood. In addition to helping to trap finer organic matter being carried down the stream, the wood itself gradually breaks apart into smaller fragments that bacteria, microbes, and stream insects can eat, before they in turn become fish food.

Large jams can so effectively obstruct stream flow that they force water over the stream banks, increasing the depth and extent of overbank flow and the amount of time that the floodplain is inundated. This extra boost to overbank flooding is wonderful for fish species that use floodplains for spawning or rearing habitat. Some of the water overtopping the banks concentrates into secondary channels that branch from and then rejoin the main channel. These smaller channels have different depths and velocities of water flow, as well as differences in water chemistry, water temperature, and streambed sediment size, all of which create diverse habitat for fish and the insects that fish eat.

In short, there are few things about wood in a stream that are not good for fish. Numerous species of fish, from large Murray cod (*Maccullochella peelii*) in the Murray River of Australia, to many species of trout and salmon, thrive in the presence of wood, and a key part of many river restoration projects designed to enhance fisheries is to get wood back into streams via natural processes or deliberate introduction by people.

Most of the advantages conferred on fish by wood and logjams are also associated with beaver dams, as well as the added benefit of off-channel beaver dams and ponds. Again, we can start with an observation by Enos Mills, who described trout surviving during a dry winter only in the deep holes of beaver ponds. It

The still waters of the beaver pond mirror the adjacent trees. A color version of this figure is included in the insert section.

took scientists a while to catch up with Mills, but eventually the studies started to come in. Along streams and rivers in California, New Mexico, Maine, Wyoming, Minnesota, British Columbia, Montana, Ontario, and even the blackwater streams of South Carolina, abundant evidence indicates that fish benefit from the activities of beavers.

The details, of course, vary between individual sites. Fundamentally, beaver dams change flowing water to still water. Accompanying this are changes in water temperature, water chemistry, sediment deposited on the streambed, the type and number of plants, microbes, and aquatic insects living in the stream and, ultimately, the fish communities. Consider water temperature, for example. Many species of fish and aquatic insects are sensitive to water temperature: either they cannot survive in water that is too cold or too hot, or their lifecycles are affected—insects that require only one growing season to mature in warmer streams may require two growing seasons before they can emerge from the stream as winged adults if the water is cold. By storing surface water in ponds, beaver dams can buffer the daily extremes of water temperature in summer. In addition, downwelling into the streambed upstream from a beaver dam and upwelling back into the channel downstream from the dam can create cool-water refuges during periods of warm air temperatures.

These refuges for fish and insects may be especially important in streams experiencing "human-induced thermal degradation," which is science-speak for streams that people have made warmer. How do we make streams warmer? Let me count the ways. We reduce the flow in the stream by diverting water for agriculture and municipal or industrial consumption. Smaller amounts of water in a channel are easier to heat up during a warm day. We cut down the riverside forest and reduce the shading that helps to keep the water cooler. We remove logs, beaver dams, and even large boulders that help to create the pressure gradients that drive water into the streambed, cooling the water before it once more wells up into the channel downstream. We release heated water from our power plants and other industrial operations. And we pump groundwater, in some places lowering the water table so much that smaller volumes of cool groundwater flow into a stream.

In small streams with cool water, beaver ponds can provide better habitat for large trout. In streams that already have warm water, numerous shallow beaver ponds in which the water grows warmer during summer can effectively decrease habitat for trout or salmon, which may be replaced by fish such as carps, minnows, suckers, perch, percids, or sunfish that prefer warmer water. The net effect is greater diversity of fish in the stream. Larger-bodied fish, such as sunfish and pike, can also displace smaller-bodied fish such as minnows in streams with warm water. Beaver ponds can provide refuges from extremes of high or low water, low oxygen, or high water temperatures in a stream, although sedimentation within the ponded area can cover spawning sites.

At the broadest level, beaver dams increase the abundance and diversity of organisms by increasing the diversity of habitats available along a stream. Working on the Kabetogama Peninsula where Bob Naiman studied beavers, for example, Isaac Schlosser and Larry Kallemeyn demonstrated that an abundant and diverse fish community requires the mosaic of habitats associated with beaver activities. Joseph Smith and Martha Mather documented similar patterns in Massachusetts. Swedish scientists studying forest streams where beavers had been recently reintroduced also found that habitat changes caused by beaver activities influenced the relative abundance of fish species and increased fish diversity. In New Brunswick, Canada, Sean Mitchell and Richard Cunjak found that beaver dams sufficiently obstruct headwater streams during late autumn base flows to limit the upstream extent of Atlantic salmon (*Salmo salar*). This allows other fish species that do not coexist with salmon to persist in the upper reaches of the stream, thus increasing the overall biodiversity of the stream. Joel Snodgrass and Gary Meffe found greater fish diversity in beaver ponds less than 17 years old located on headwater streams in South Carolina, but lower diversity in older ponds. They interpreted these results as indicating that beavers positively affect fish biodiversity most effectively when beavers remain in a stream, creating new ponds even as they abandon old ponds.

Reviewing studies from across North America, Paul Kemp and his colleagues found that the most frequently cited benefits of beaver dams for fish are increased habitat heterogeneity, rearing and overwintering habitat, refuges during high and low flows, and increased production of food in the form of insects. The most commonly cited problems for fish are dams impeding fish movement, siltation of spawning habitat, and lower levels of dissolved oxygen in ponds. Descriptions of benefits greatly outweighed descriptions of problems, however, and the consensus of fisheries experts is that beavers positively impact fish populations because of their influence on habitats and insect abundance.

The diversity of aquatic insects in a stream also increases with habitat diversity. In small woodland streams in Quebec, beaver ponds provide quiet water habitat that hosts different aquatic insects than are present in the flowing stream. D. M. McDowell and Bob Naiman demonstrated that the ponds collect up to 10 times more fine organic matter than riffles in the main channel, and the total amount of insect flesh (otherwise known as biomass) per square foot in the ponds is 2 to 5 times greater than in the riffles during the active seasons of spring and summer. Among other things, this means more food for fish, frogs, songbirds, spiders, and other predators of insects.

One of the fascinating recent developments in ecology is using naturally occurring isotopes in the tissues of plants and animals to identify which plants the animals are eating. Aquatic and terrestrial plants contain different ratios of carbon isotopes. In an ingenious study, Magnus McCaffery and Lisa Eby analyzed the carbon isotopic ratios of aquatic mosses, macrophytes, and filamentous algae in streams of southwestern Montana, as well as the isotopic ratios of willows, grasses, and other terrestrial vegetation along transects extending away from each channel. The researchers then measured emergence rates of aquatic insects such as mayflies, along with the abundance and tissue-isotopic composition of wolf spiders (Lycosidae) (wolf spiders were chosen because they prey on emergent aquatic insects).

The intent of the study was to measure aquatic subsidies to terrestrial consumers. This sounds vaguely like a government providing extra water to shoppers on land, but ecologists use the word "subsidy" to describe transfers of nutrients and energy between different components of the environment. A salmon spending its adult life in the ocean, eating ocean food and growing large, migrates up a river to spawn and die. The dead salmon provides a marine subsidy to the terrestrial organisms such as bears or eagles that feed on its tissues. Riparian trees drop leaves and twigs into a stream, providing a terrestrial subsidy to the stream ecosystem. The aquatic insects that emerge from the stream as adults provide a stream subsidy to the adjacent terrestrial environment if the insects are eaten by spiders, birds, and other terrestrial animals.

None of this should be surprising, but people sometimes like to compartmentalize the world a little too much, and it took ecologists a while to realize that a stream is not an isolated entity but is instead intricately connected to the ocean, the atmosphere, and the land via numerous pathways of matter, energy, and organisms. Once these exchanges are recognized, the scientific challenge becomes how to measure them. As the old saying has it, you are what you eat—and what you eat is recorded in the isotopic ratios of your body.

In the Montana study, McCaffery and Eby compared streams with beaver activity to those without and found two important differences. First, sites with beaver activity had more food, with more than 200% higher aquatic insect emergence rates. Second, the beaver sites had more eager eaters of that food, with 60% higher abundance of spiders. The tissues of the spiders contained a greater percentage of carbon derived from aquatic environments. This indicates the food chain developing as a result of the activities of beavers: more habitat for aquatic plants results in more aquatic plants, which are eaten by aquatic insects, which in turn feed spiders.

In another study, Matthew Fuller and Barbara Peckarsky found that larger beaver ponds in creeks of the Colorado Rockies result in larger female mayflies (*Baetis bicaudatus*). Because larger females are more fecund, the beaver ponds likely indirectly cause more mayflies, which again translate to more food for mayfly predators such as spiders and songbirds.

A comparison of a freely flowing reach of a small stream in Germany with a reach containing a beaver dam and a beaver pond demonstrated that even the dam itself provided good habitat—insect diversity and abundance were higher in the dam than in the freely flowing stream reach. Beavers in Alberta that increased the extent of open water during drought also promoted greater wetland biodiversity by creating unique niches and microhabitats for insects, especially during dry periods.

Along a stream in Missouri, investigators found that beaver ponds host five times the amount of plankton—floating algae—as a stream without beavers. The ability of the ponds to trap silt and clay increases habitat diversity for bottom-dwelling invertebrates. Although standing water is commonly considered to be good mosquito-breeding grounds, mosquitoes were less numerous in the beaver ponds, perhaps because the ponds also support mosquito predators. The ponds host more fish species and more individual fish, as well as providing habitat for salamanders, frogs, and toads.

Salamanders and frogs, in particular, are not faring well in this world that humans have so heavily modified. As many as 112 species of amphibians went extinct between 1980 and 2006. Populations of remaining species are declining swiftly. In tropical regions, rising temperatures enhance the spread of the chytrid fungus deadly to frogs. Elsewhere, the combined effects of deforestation, pollution, and loss of wetlands are the immediate causes of extinction, but behind all of these looms the enormous shadow of humanity.

One of the challenges in understanding the decline of any population or species is to identify which of the many threats to survival can make the difference between persistence and extinction. The Columbia spotted frog (*Rana luteiventris*) is among many species of frogs in the western United States that is in trouble. Scientists Robert Arkle and David Pilliod identified year-round aquatic habitat with emergent vegetation such as rushes and non-trout fish species as one of the critical needs of a genetically distinct portion of the frog population living within the Great Basin of Nevada and southern Idaho and Oregon. And—beavers to the rescue—streams occupied by beavers were most likely to include such frog habitat. Arkle and Pilliod found that Columbia spotted frogs were much more likely to be present in active beaver ponds than in water bodies without beaver or inactive beaver ponds, indicating that both the ecosystem and its engineers are critical to frog survival. The importance to frog survival of beaver engineering is only likely to increase in this already-dry region in which ponds may become rarer as climate warms.

Sadly, protecting lands in national parks is not enough to ensure frog survival. Blake Hossack and others studied five wetland-breeding amphibian species in Glacier, Yellowstone, Grand Teton, and Rocky Mountain National Parks. During the study period of 2002 to 2011, the scientists found more declines than increases in frog populations, but they also found that beaver meadows supported more amphibians for four of the five species monitored. Boreal toads (*Anaxyrus boreas*) and Columbia spotted frogs were particularly fond of beaver meadows.

Beaver activities enhance frog survival in wetter climates, too. The canals that beavers excavate across a floodplain are used by pond-breeding amphibians during dispersal and migration between aquatic and terrestrial habitat. Working in Alberta, Canada, Nils Anderson and his colleagues found as many as nine times more wood frogs (*Lithobates sylvaticus*) in beaver canals than along pond shorelines not modified by beavers. Adult wood frogs live in the beaver canals and juveniles migrate along the canals to new homes. Reading that study made me more tolerant of the canals that I so frequently step into and stumble over. From now on, I'll think of them as frog highways.

Slightly farther south, in Upstate New York, the frog highways may be even more important. Mink frogs (*Rana septentrionalis*) living there are at the southern limits of the species' range, mainly because mink frogs are sensitive to both warm temperatures during the July breeding season and desiccation. Mink frogs mostly live in active beaver ponds that are least 3.7 acres in size. Predicted warming and drying trends in the region have the potential to eliminate the frogs from Upstate New York, but the presence—and particularly an expanded presence—of beavers in the area may provide the habitat and the dispersal corridors necessary to sustain a viable population of mink frogs.

The Columbia spotted frogs in the Great Basin and the mink frogs in Upstate New York are both at the southern, and therefore warmest and driest, edge of their species' geographic range. It is species on the edge that are most at risk from climate change and other forms of habitat loss and for which beaver ecosystem engineering is vital to survival. Fish are certainly among these species.

Across North America, scientists have documented more than 80 fish species using beaver ponds. All the familiar types—perch, suckers, pickerel, darters, shiners, cutthroat trout, brown trout, brook trout, minnows, sculpins, pike, bass, bullheads, salmon, chubs, and bluegill—use the habitat engineered by beavers. Of these, 48 fish species commonly use beaver ponds, which have slower water velocity, extensive

A wood frog in the North St. Vrain beaver meadow. A color version of this figure is included in the insert section.

cover for fish, a productive environment for vegetation and aquatic insects, and a large expanse of the water's-edge habitat that many fish prefer. Fish use less energy when foraging in slow water, so beaver ponds typically contain more fish and larger fish than adjacent stream habitat. Ponds provide good overwinter habitat. Although the mucky bed of a beaver pond isn't good for spawning gravels, the presence of a pond can slow the passage of a flood down a stream, reducing siltation and erosion of spawning gravels in the channel. Even fish species such as brook and cutthroat trout, which require gravels for spawning, can use beaver ponds as rearing habitat. And, by damming very small, steep streams or seeps from valley side slopes along a larger river, beavers can create fish habitat where none previously existed.

Of course, as fish populations rise, so may predation by piscivores. Mink, otters, and birds, including hooded mergansers (*Lophodytes cucullatus*), green-backed herons (*Butorides striatus*), great blue herons (*Ardea herodias*), and belted kingfishers (*Ceryle alcyon*), are more abundant in wetlands created by beavers than elsewhere along a stream.

In a warming and drying world, the most important effect of beaver engineering may be increasing resilience to drought, just as Enos Mills described more than a century ago. The Yurok Tribal Fisheries Program has been restoring salmonid habitat on the lower Klamath River in northern California, in part by building stick dams that mimic those of beavers. Fish biologists working for the program have found that only in beaver ponds have endangered Coho salmon been able to survive four years of below-average rainfall. Brock Dolman, a leader of California's Bring Back the Beaver Campaign, refers to loss of beavers as a "double whammy for a watershed" because of the increased flooding, channel erosion, sediment deposition, and groundwater recharge during wet periods and the reduced stream flow during drier periods.

So, after many decades of research, attitudes toward beavers have come full circle in the context of fish management: if you want more fish and fish food, let the ecosystem engineers get to work.

Cutthroat Competition

A brook trout moves slowly into the nearby shallows as I sit quietly beside the pond in the North St. Vrain beaver meadow. It is a beautiful fish, streamlined and moving with silent agility, the thin white bands edging its fins easily visible against the dark bed of the pond. Brook (*Salvelinus fontinalis*) and rainbow (*Oncorhynchus mykiss*) trout, native to the eastern United States and the U.S. West Coast, respectively, are now the most common trout species in streams of the Colorado Rocky Mountains. Repeatedly introduced by fish biologists, private angling clubs, and individual anglers, these species have outcompeted the native greenback cutthroat trout (*Oncorhynchus clarkii stomias*), which is now endangered. The combined effects of 19th-century mining, timber harvest, flow regulation, and overfishing dealt cutthroat populations a severe blow. Introduction of brook and rainbow trout then largely finished the process of eradicating the natives. Greenbacks are now present in only a few headwater stream segments protected by the combined effects of aggressive fisheries management and a tall waterfall that prevents upstream migration by brook and rainbow trout.

I watch the brook trout move leisurely back toward the deeper part of the pond. Historical photos show anglers posing in front of masses of fish caught in the streams of Rocky Mountain National Park and sold in the fish markets of Denver. When anglers started systematically working these streams in the 1860s and 1870s, it was possible to catch as many as 60 trout in a single large pool. Such abundance is unimaginable today, but the presence of at least some functioning beaver meadows undoubtedly helps to maintain the fish populations still present.

Work by my colleague Kurt Fausch demonstrates that trading one species of trout for another can have subtle but profound effects on a river ecosystem. Think of a river as a flow of energy. Water is only the most obvious component of this flow. Energy flows into the river as sunlight that powers photosynthesis and as food in the form of leaves, pine needles, and twigs dropped from riparian forests. Microbes and bottom-dwelling stream insects such as caddisfly and mayfly larvae graze on algae and the litter from dead plants. When the insects emerge as winged adults, energy flows back to the forest in the form of insect bodies consumed by spiders and birds. Just as water flowing down a river swirls in eddies and rushes through standing waves rather than flowing with uniform smoothness, so the energy flowing along a river can sometimes follow complicated pathways. Differences among trout species create some of these complexities. Cutthroats mostly capture insects drifting on the water surface. Brook trout feed on bottom-dwelling grazing insects. By reducing these streambed grazers, brook-trout feeding can cause an increase in streambed algae, which in turn affects insect communities and everything that feeds on them. Brook trout can also reduce the number of adult insects emerging from the stream by more than half, which likely has a greater effect on other insect predators, from ouzels to spiders to songbirds.

At the beaver pond along North St. Vrain Creek, feeding trout create patterns of widening circles of ripples as the sky clouds over again. The process is quiet and I imagine a swift, controlled rush upward that culminates in trout lips delicately plucking an unsuspecting insect from the slender boundary between air and water. As with every other aspect of this meadow, the activities of the beavers support the feeding trout and the anglers who come to match wits with the fish.

August

Legacy Effects

Emily Dickinson wrote a lovely poem using a brook as a metaphor for one's interior life. The poem includes the lines:

> And later, in August it may be,
> When the meadows parching lie,
> Beware lest this little brook of life
> Some burning noon go dry!

No chance of the little brook going dry if it runs through a beaver meadow.

The movement of water across and through the North St. Vrain beaver meadow has slowed perceptibly by August. Some of the secondary channels barely flow and the main channel is easily crossed on foot. The water remains high in the main beaver pond, but few of the small dams winding across the meadow have water spilling over them. My feet are less likely to sink into wet black muck as I wander through the meadow, and even the moose tracks leave less of an imprint in the drying soil. Plenty of water remains, however, and the meadow is a much brighter shade of green than the adjacent, drier hill slopes.

Many flowers remain in bloom across the meadow. Stalks bristling with the elaborate, richly pink blossoms of elephant's head rise above standing water. Dusky purple monkshood flowers in slightly drier soil, as do the showy blue and white columbines. The blue bell-shaped flowers of harebell mark the driest sites. The late-summer flowers are joined now by the spreading tan or scarlet caps of fungi, as well as green berries on the ground juniper and kinnikinnick growing on the drier terrace beside the beaver meadow. Songbirds born this summer are fully feathered and capable fliers, and some of the birds have already left the meadow for the year. Early morning temperatures carry a hint of the coming autumn.

A portion of the North St. Vrain beaver meadow in August: not quite parching, as wetland plants encroach on the steadily retreating standing water. A color version of this figure is included in the insert section.

The beaver kits grow steadily more capable, too, and by now they are used to foraging on their own. Presumably, this frees the breeding adult female for more time spent in dam and lodge repair or starting the food cache for the coming winter.

I still do not see the North St. Vrain beavers, but I am able to watch a pair of beavers along the Fall River, a few drainages to the north. The Fall River in Horseshoe Park is another of the formerly active beaver meadows in Rocky Mountain National Park that is now largely abandoned. Willows are now restricted to a narrow band along the river channel, with grasses and low woody shrubs growing across the rest of the valley bottom. The Fall River has abandoned most of its secondary channels and cut downward to flow between tall, vertical banks. In a large, fenced, grazing exclosure at the downstream end of Horseshoe Park, a pair of intrepid beavers have recently recolonized the site. The beavers have built a den into the stream bank, with a sub-stantial pile of branches and woody stems piled over the den. Sometimes known as "bank beavers," these beavers are the same species as those that build dams. Where a channel has cut down too deeply or is simply too large to dam, beavers can still live along the channel but will build a bank den rather than a lodge in the middle of the pond.

Watching the Fall River beavers is both exciting and frustrating. Exciting, because I actually get to see the beavers rather than just seeing the spreading circles from a tail slap and a beaver's dive. Frustrating, because the animals remain so elusive. I sit quietly on the bank opposite the beaver den, my feet in the cold water and the rest of me sweating on this hot August afternoon. I stare intently at the spot that I expect to be the entrance to the den, unwilling to look away and miss a glimpse of the beavers. The minutes pass. Mosquitoes whine and then bite. Hummingbirds buzz down to investigate me and then whir away, disappointed at my lack of nectar. Then, abruptly, I see a darker shadow at the base of the opposite bank. For a moment, I think it is a fish or even a shadow caused by turbulence in the water, but the shadow is too wide and brown and—it has a broad, flat tail! The beaver never surfaces but instead follows the edge of the vertical bank upstream like someone tracing a line. Where the bank is undercut at the water's edge, the beaver vanishes beneath the overhang and continues upstream, without a sound or a ripple. I continue to sit unmoving and am rewarded by the sight of a second beaver following the first. The two come and go from the den entrance over the next half hour without ever rising above the water surface or making the slightest sound. Even a few moments' inattention during one of their brief passages and I would miss seeing them.

Creating a Fur Desert

Although humans have presumably never threatened these particular beavers in Rocky Mountain National Park, the beavers have good reason to remain cautious. Frustrated as I am by my inability to see the beavers more easily, I realize how wonderful it is that beavers remain here at all. Numerous books have examined the history of the fur trade, which was integral to European exploration of North America. Europeans were delighted to be able to trap beavers in North America because European beavers had been driven nearly to extinction by the fur trade. Royal edicts were needed to protect beavers in Europe by the 18th century because of the 250-year-long demand for beaver fur to be used in men's hats. Beaver fur is uniquely able to hold its shape through the process of felting. This characteristic allowed fashion designers to use beaver fur for fine-quality, wide-brimmed hats such as those worn by the Three Musketeers and tall hats such as the cocked hat of naval officers or the various styles of hats that resembled stovepipes. If 19th-century women were largely responsible for the demise of many bird populations because of the demand for bird feathers or occasionally whole birds in women's hats, men of the 17th, 18th, and early 19th centuries were largely responsible for the demise of beaver populations across Eurasia and North America.

In North America, Native Americans also valued the warmth and beauty of beaver fur, and they hunted the animals prior to European colonization of the continent. In his book *Keepers of the Game*, Calvin Martin traces how the relationship of Native Americans to beavers changed through time. Cultural taboos against overhunting allowed beaver populations to thrive prior to European contact, but establishment of a commercial fur trade in which Native Americans participated led to local extinctions of beaver populations and then massive hemorrhaging of beaver numbers across the United States and southern Canada.

Despite the occasional protective royal edict, European beaver populations had reached extremely low levels by 1638, when fur traders set their sights on North America. In *The Beaver Manifesto*, Glynnis Hood notes that Jacques Cartier's 1542 trip up the St. Lawrence River ended at a place the Native Americans called Hochelga, or beaver meadows; we now call it Montreal. In *Fur, Fortune, and Empire*, Eric Dolin describes the importance of beaver furs as a commodity that could be exported to Europe, allowing the early English colonists in North America to purchase from Europe the many products that they could not provide for themselves, including such basics as food and clothing. The European presence in North America was built on the dead bodies of millions of beavers.

The westward-spreading fur trade decimated beaver populations across North America. Jesuits reported beavers disappearing from the Three Rivers area near Quebec as early as 1635. The number of beavers in what is now New York City declined by 1687, and by 1700 the fur trade in the area was largely finished. Eastern Wisconsin was "overharvested" by 1740, and by 1790 beavers were extinct throughout Wisconsin. Hood describes how the Hudson's Bay Company, newly merged with the North-West Company in 1821, deliberately set out to eradicate beavers in Oregon Territory in order to create a "fur desert" that would limit rival American traders. This short-sighted greed resulted in extirpation of beavers from much of the Pacific Northwest in less than 20 years, but it was hardly unique. Beavers were almost extinct throughout North America by 1900. In some regions, they have yet to return.

Much has been written about the fur trade. Scholars have explored the history of trapping and commercial trading, the methods used to kill beavers, the socioeconomic and political implications of the trade for Native American, Euro-American, and European societies, and the environmental implications, among other things. Little has been written about the effect on beavers as individual animals. As herbivores, beavers are subject to predation by various carnivores, and beavers die through starvation and other natural processes. Under natural conditions, however, it seems less likely that all of the beaver colonies in an area would be entirely destroyed, leaving no beavers. Bernie Krause has been recording natural sounds since 1968 and has founded a field known as soundscape ecology. In a 2013 TED talk, Krause described "probably the saddest sound I've heard coming from any organism, human or other": the sound of a lone male beaver swimming slowly in circles for hours, crying for his lost mate and offspring after a pair of game wardens had dynamited a beaver dam and killed the female beaver and her kits.

Colorado Mountain Men

The history of beaver trapping in the region of Rocky Mountain National Park was no different than the rest of the continent. The Intermountain West was spared from commercial beaver trapping until after the Louisiana Purchase in 1803 and the Lewis and Clark expedition during May 1804 to September 1806. Although beaver trapping was technically illegal in the region in the early 19th century, historians believe that illegal commercial trapping occurred in the Southern Rockies before 1821. Trapping is better documented during the next two decades, when men such as Ceran St. Vrain (1802–1870) were active in the region. The Big and Little Thompson Rivers just to the north of North St. Vrain Creek were likely named for Phillip Thompson, a fur trader in the region during the early 1830s. In 1831, American fur brigades began to trap in and around North Park and Middle Park, broad, high-elevation basins to the northwest and west of Rocky Mountain National Park.

Ceran St. Vrain was born in Missouri, of French ancestry. He worked with the trapper-trader brothers Charles and William Bent from 1831 to about 1850 and lived in the Southern Rockies from 1824 to 1870, but most of his trapping and fur-trading activities occurred during the later 1820s and 1830s. He traveled to North Park, within the mountains north of Rocky Mountain National Park, in 1827. By 1837 or 1838 he built a trading post known as Fort St. Vrain at the confluence of St. Vrain Creek and the South Platte River. The trading post lasted for about 6.5 years but was abandoned by 1844.

The abandonment of Fort St. Vrain likely at least partly reflected the increasing scarcity of beavers. Comparing the earliest written descriptions of the region reveals the swift decline in beaver populations. Edwin James, chronicler of the 1819–1820 expedition led by Major Stephen Long, described the behavior of the beavers that "abounded" in the upper branches of the Platte River drainage:

> Three Beavers were seen cutting down a large cotton wood tree: when they had made considerable progress, one of them retired to a short distance, and took his station in the water, looking steadfastly at the top of the tree. As soon as he perceived the top begin to move towards its fall, he gave notice of the danger to his companions, who were still at work, gnawing at its base, by slapping his tail upon the surface of the water, and they immediately ran from the tree, out of harm's way. (p. 464)

By the time John Charles Frémont traveled through the area around Rocky Mountain National Park in 1842, he observed numerous abandoned beaver dams and lodges, but almost no active colonies:

> We . . . encamped on a pretty stream, where there were several beaver dams, and many trees recently cut down by the beaver. We gave to this the name of Beaver Dam creek, as now they are becoming sufficiently rare to distinguish by their name the streams on which they are found. In this mountain they occurred more abundantly than elsewhere in all our journey, in which their vestiges had been scarcely seen. (Frémont, 1845, p. 352)

The collapse of the beaver fur market when silk hats became fashionable in Europe during the late 1830s also helped the few remaining beaver in North America. Some beavers remained in the area, however, because in 1913 Enos Mills wrote of trappers killing more than 100 beavers in the Moraine Colony, just north of the North St. Vrain beaver meadow.

Although few in number, settlers attempting to farm in the vicinity of what later became Rocky Mountain National Park could also strongly affect beaver populations because both farmers and beavers preferred the same habitat. Describing homesteading efforts along the upper Colorado River valley on the western side of the national park during the 1880s, Thomas Andrews writes in his book *Coyote Valley* of the efforts of settlers such as Sam Stone. Stone dug a channel more than 1,500 feet long, 2 feet wide, and 3 feet deep through Big Meadows in order to drain the wet meadows for cultivating hay. Clearing of willows and other native riparian vegetation, along with draining the meadows, drove the beavers out. Thomas describes Clinton DeWitt's beaver battles in the Upper Colorado during the 1920s, which involved destroying beaver dams and then killing beavers. DeWitt wrote that beavers "so infested the valley on which my homestead lies . . . that practically all of the hay land is continually flooded and cut up by beaver runs" (Thomas, p. 223). On the eastern side of the national park, Peter Hondius changed Upper Beaver Meadows into hayfields.

Despite these relatively late trapping ventures and some hay cultivation, beaver populations in Rocky Mountain National Park did recover during the 20th century. Censuses of beaver populations in and around the national park have been conducted irregularly throughout the 20th century, starting in 1926 and continuing to the present. The documents resulting from these surveys can be challenging to interpret because different geographic locations were surveyed through time, and it is not clear whether the absence of a particular location reflects an absence of beaver at that site or simply a failure to inventory existing beaver populations at the site. Most of these surveys do not include the North St. Vrain Creek drainage. A 1947 report by Fred Packard notes 68 beavers in the North St. Vrain drainage within the national park, at a time when more than 600 beavers inhabited the western portion of the national park. Packard wrote that "Almost every stream that can support beavers . . . [is] stocked to capacity or . . . overpopulated" (Thomas, p. 230), and Park Service rangers collaborated with ranchers to remove beaver dams and kill beavers.

At present, no beavers remain on the western side of the park. A 1999 survey notes two active beaver areas along North St. Vrain, whereas a 2000 survey reports seven active sites within the drainage, but neither of these later reports includes a population estimate. If the seven sites active in 2000 had about 6 beavers per site, this would work out to about 40 beavers in the watershed. I'm reasonably sure that fewer than 40 beavers live within the main beaver meadow at present, but I do not know how many are there.

U.S. Geological Survey biologist Bruce Baker noted in 2005 that beaver populations in the national park declined dramatically in the 1940s and then failed to recover. As part of an attempt to understand why, Baker drilled small holes in the tails of 41 beavers and attached radio transmitters that would allow the movements of the animals to be tracked. This technique was adopted following earlier failures by other researchers who clearly demonstrated that beavers are difficult to tag. Early attempts included neck collars, but "beavers have fusiform bodies with thick necks that cause collars to fall off" (Baker, 2005, p. 218). Scientists tried implanting transmitters, but this requires surgery and can kill the beaver. Collars placed around the base of the tail can also fall off if the tail decreases in size following attachment. Unfortunately, nearly all of the transmitters attached to holes in the beavers' tails became detached, some of them through being chewed off, so the attempt was a failure. Beavers preserve their privacy.

When David Mitchell, Jennifer Tjornehoj, and Bruce Baker surveyed beaver populations in the national park during 1999, they found fresh signs of beaver in the form of two lodges, dams, fresh cuttings, and a food cache along North St. Vrain Creek. Ouzel, Sandbeach, and Hunter's Creeks showed signs of only past beaver activity. Similarly, the Glacier Creek drainage showed only past beaver activity. However, the upper Poudre River, upper Colorado River, Cow Creek, Fall River, and Big Thompson River catchments all showed signs of contemporary beaver activity. When I visited these valleys in 2011, only Cow Creek and portions of the Big Thompson River drainage still had a few beavers, although by 2017 the pair of beavers had moved into the Fall River site. The legacy of the early Colorado mountain men and the settlers who followed them continues, likely abetted by the large numbers of elk and moose now present within the national park and competing with the beavers for food supplied by riparian vegetation.

A Legacy of Absence

Ecologists and geomorphologists use *legacy* to describe features of a landscape or ecosystem that reflect past human activities. Legacy sediments fill valley bottoms across the United States, a reminder of mill dams now long abandoned and forgotten, of uplands cleared for crops and then left to regrow into forest, or of placer metals blasted from mountain valleys using destructive, 19th-century hydraulic

A beaver harvesting willow branches along the Dall River in central Alaska. A color version of this figure is included in the insert section.

mining techniques. The species of aquatic insects present in a stream may be a legacy of agriculture that occurred half a century earlier, or the presence of fish in a high mountain lake may be a legacy of fish introduced by anglers a century ago.

Human destruction of beaver populations throughout the Northern Hemisphere created another kind of legacy, a legacy of absence. In the absence of beaver dams, valley bottoms dried and became less diverse, rapidly passing water, sediment, and nutrients downstream, and supporting fewer and less diverse plants and animals. The changes are so ubiquitous and lasting that it is only when I visit a relatively natural site such as the North St. Vrain beaver meadow that I become aware of this legacy.

My PhD student DeAnna Laurel quantified the effects of this legacy of absence on stream flow, spatial heterogeneity, and storage of organic carbon in beaver meadows. Written records and air photographs allow us to constrain the time of abandonment for many of the beaver meadows in Rocky Mountain National Park. DeAnna divided her study sites into three categories of active, recently abandoned (within the past 20 years), and long abandoned (for more than 30 years). She found that beaver legacies persist for at least two decades after the animals abandon a meadow. The recently abandoned meadows are not significantly different than active meadows with respect to the surface spatial heterogeneity of vegetation and topography and

Beyond the edge of the drier terrace, the beaver meadow remains wet and green in August. A color version of this figure is included in the insert section.

the subsurface characteristic of soil moisture. Even the long-abandoned meadows do not differ significantly from the active meadows in terms of soil depth and soil organic carbon content. The good news is that, although habitat abundance and diversity decline with time following beaver abandonment, this decline is not immediate and can be reversed if beavers return to the site. We just need to allow them to return and, where necessary, foster the conditions that make it possible for them to survive.

September

Alternate Realities

The first week of September mostly feels like summer. The air on the dry terrace bordering the beaver meadow is richly scented with pine. Purple aster, blue harebells, and tall, yellow black-eyed Susan still bloom. Fungi are more abundant on the forest floor, and the tiny, purplish berries of kinnikinnick are sweet to the taste. The air is warm in the sunshine, but strong winds hurry rain showers through at intervals. Patches of last year's snow linger on the surrounding peaks, even as the first light snows have already fallen in the high country.

Down in the beaver meadow, the leaves of aspen, willow, birch, and alder are starting to assume their autumn colors. Here and there a small patch of yellow or orange appears among the green. Blades of grass have a pale orange tint and the strawberry leaves have gone scarlet, even as white asters, purple thistles, and a few other flowers continue to bloom. The creek is noticeably lower, its cobble bed slick with rust-brown algae. Exposed cobble and sandbars have grown wider as the water has shrunk back from the edge of the willows, and the main channel is easy to cross on foot. The clear water is chillingly cold in both the main channel and the side channels. The smaller side channels no longer flow, and a drape of mud mixed with bits of plants covers the cobbles. Wood deposited a year ago has weathered to pale gray. The older, marginal beaver ponds have shrunk noticeably, and the water is lower in the main ponds, where tall sedges now lie bent on the top of the declining water surface. The beavers remain active: following fresh moose tracks, I come on a newly built beaver dam on a small side channel.

By the third week of September, autumn has clearly arrived in the mountains. The air remains quite warm during the day, but nights of frost are swiftly bringing out the autumn colors. Whole stands of willows and aspen now glow golden or pumpkin-orange. Flow in the creek is very low and I see beaver food caches in most of the ponds. A flock of warblers chirps from the willows beside the main channel, but they are migrants on the move, stopping only briefly as they fly south to warmer regions.

Sedges bent partway down the stalk as the water level in the beaver pond declines.

A beaver food cache exposed along the edges of a shrinking pond.

The beautiful autumn colors appearing among the foliage of the beaver meadow are a reminder of life retrenching as photosynthesis shuts down for the season. Abert's squirrels are as active as usual, but their mounds of dismantled cones at the base of large conifers show evidence of recent disturbance—the broad, spongy mass is riddled with small holes where the squirrels are caching winter supplies. I imagine the beaver kits wondering at the changes in their watery world. Some of the favored summer herbaceous foods are dying back and the adults and yearlings cache food for winter.

I see some of this activity at the main pond, where one of the beavers is swimming from the shore into deeper water to cache a stripped branch. The beaver's movements create only a small wake in the still water. If the beaver was not carrying the white branch in its mouth, its head protrudes so little above the surface that it would be difficult to detect from a distance. The animal's sense organs are aligned in a row so that a beaver swims with nostrils, eyes, and ears raised above the water while the rest of the head and body remain submerged. A slight movement on my part and the beaver is gone, not even bothering to slap its tail as a warning to others in the colony.

The daily activities of the beavers now cover a smaller area than during summer, although their home range will shrink to its smallest extent during winter. I never get a sufficiently clear view of these wary animals to assess whether they look fatter

A swimming beaver pushing a log.

or leaner, but I imagine them putting on weight against the long, cold winter. I sit for a while beside the beaver pond, enjoying the subtler, background sounds of bird calls and breezes moving through the trees now that the roar of the creek has subsided to the murmur of low flow.

Beaver Meadows and Elk Grasslands

Elsewhere in the national park late September can be a noisy time, with the weird whistlings and squeals of rutting elk (*Cervus elaphus*) echoing through the valleys. The accepted term for these sounds is bugling, but they do not resemble any bugle I have ever heard. The male elk are magnificent in appearance at this time of year, with finely polished, sharp-tipped antlers, muscular chests and necks, and thick, smooth coats, but I can only laugh at the ludicrously high-pitched sounds that emerge when they open their mouths. The female elk find the squeals appealing, however, and that is all that matters. People travel up from the urban areas at the base of the mountains and from much greater distances to enjoy the autumn show of golden aspen leaves and vocal elk. The elk are arguably the single most charismatic megafauna in this park. Coyotes, moose, and black bears can be seen, but much less frequently. Mule deer are also present, but are more common outside the park. Others once present—wolves, grizzly bears, bison, wolverines, lynx—are extinct here. Mountain lions are so rare and elusive that only the very lucky catch a glimpse.

Elk, on the other hand, are ubiquitous. They cause traffic jams, either by crossing the road or, much more frequently, simply by being visible from the road. Elk grazing along the Big Thompson River in Moraine Park, the peaks of the Continental Divide forming a spectacular backdrop, is one of the iconic scenes of Rocky Mountain National Park. Some problems underlie this iconic scene, however.

As with many other large forms of wildlife, people hunted elk nearly to extinction by the late 1800s in what later became Rocky Mountain National Park. Elk were reintroduced to the national park during 1913–1914, when 49 elk were transplanted from the Yellowstone herd. By 1940 the descendants of these pioneers had increased to about 1,200 animals, at which time the Park Service undertook control efforts that reduced the elk population to about 500 until 1968. Discontinuation of these efforts allowed elk numbers to reach 3,000 by the late 1990s. By then, elk numbers and what to do about them had become quite controversial. Culling—a polite word for deliberate shooting to reduce numbers—was unpopular with the public, even if done in a low-visibility manner by professional hunters. Eventually the Park Service settled on bulky contraceptive collars worn around the neck, so that it appears as though a large portion of the female elk population has suffered from whiplash.

Similarly, people hunted moose (*Alces alces*) to extinction prior to 1900. The Colorado Department of Wildlife brought 12 moose from Utah to the western slope

Elk grazing in Moraine Park during 2011. A color version of this figure is included in the insert section.

of the Never Summer Mountains outside of Rocky Mountain National Park in 1978, then introduced another 12 from Wyoming in 1979. By 1986–1987, Colorado had an estimated 100 to 130 moose. Many of them made the easy journeys into both the eastern and western sides of the national park. The first recorded moose sighting occurred in 1980 along Onahu Creek on the western side of the park.

Although hunted in adjacent national forests, the moose population in the national park has grown steadily. The Park Service is currently undertaking a moose population census. This is more difficult than you might imagine in a park that is 98% wilderness, which prevents the use of helicopters or planes for the type of aerial census commonly used to estimate moose populations. Forty years after the first moose sighting in the park, the evidence of moose browsing is abundant, particularly on the western side of the national park, where I commonly find willows browsed to knee-high remnants in abandoned beaver meadows that are crisscrossed by moose tracks. I seldom visit the North St. Vrain beaver meadow without seeing at least one moose.

Increasing moose numbers in the North St. Vrain beaver meadow are not simply an interesting example of natural competition among herbivores. If the moose eventually increase in number sufficiently to cause the type of intense browsing pressure associated with elk in Moraine Park or Upper Beaver Meadows, food supplies may run out for the beaver colonies along North St. Vrain Creek. North St. Vrain is the last stand for beavers in Rocky Mountain National Park.

While elk and moose numbers within the park have steadily increased; beaver populations have steadily declined. Enos Mills published the earliest population estimates for selected beaver colonies. Other reports followed sporadically through time—1926, 1940, 1964, 1980, 1999, and 2000. Only the 1940 and 1999–2000 surveys included North St. Vrain Creek, but the Big Thompson River was more frequently surveyed for beavers. Beaver numbers along the Big Thompson declined from an estimated 315 animals in 1940 to none by 2000. In contrast, elk numbers for the entire national park exhibit two periods of steady rise interrupted by a period of population control between about 1944 and 1979.

The trends in the lines representing elk and beaver numbers are related. Left to their own devices, elk tend to congregate along streams with water and abundant willows, aspen, river birch, and alders to browse. Elk find these plants so succulent that they eat them to the ground, leaving nothing for a hungry beaver or a beaver intent on building a lodge or dam. Fences built to exclude elk along Beaver Brook in the central part of Rocky Mountain National Park during the 1930s and 1940s had up to 70% willow cover in 1959, whereas areas outside the fences had no willow cover.

Bruce Baker, the U.S. Geological Survey scientist who attempted to track beavers in the North St. Vrain beaver meadow with tail-mounted transmitters, worked with several colleagues to examine the effect of beaver cutting and elk browsing of willows in Moraine and Horseshoe Parks within Rocky Mountain National Park. These sites

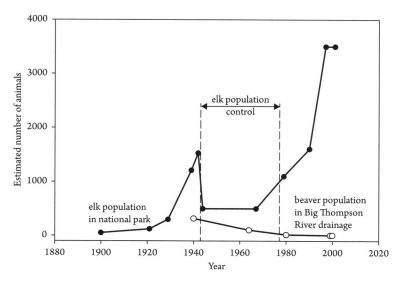

Changes in elk and beaver numbers within Rocky Mountain National Park through time.

have thickets of willow (*Salix monticola, S. geyeriana, S. planifolia*) and alder (*Alnus tenuifolia*) with an herbaceous understory of *Carex* and grass. Willows are versatile and adaptive plants: they must be, to survive in the challenging environment along streams, where a high flow can submerge the plant in water or scour away the soil in which it is rooted. Willows have adapted to resprout from a stump when the plant is sheared off near the ground. Willows can also reproduce both sexually via seeds and asexually from severed branches or new sprouts from spreading roots.

When Baker and his colleagues simulated both beaver cutting and elk browsing, the willows became short and narrow, with far fewer, but longer, shoots and a higher percentage of dead branches. When the scientists simulated beaver cutting without elk browsing for three years, they created tall, leafy willows with a total mass of stems and leaves that was 10 times greater on unbrowsed plants than on browsed plants. The scientists also found that unbrowsed plants recovered almost all (84%) of their precut mass after two growing seasons, whereas browsed plants recovered only 6% of their precut mass.

These experiments dramatically demonstrated that willows respond differently to herbivory, depending on how the plants are cut or browsed. Large mammals such as elk and moose repeatedly browse shoot tissue from growth areas at the tops of woody plants. This can cause the plants to transfer stored carbon from roots to shoots and increase nitrogen levels in leaves. Transfer of nutrients increases rates of photosynthesis and growth and causes more branching. Heavy browsing can reduce or eliminate asexual reproduction in willows and keep the plants in a juvenile growth phase that reduces plant fitness.

Beavers, in contrast, cut and remove entire stems at or near the ground surface. Beavers commonly cut all stems from preferred shrubs growing near their winter food caches, dams, and lodges, but then become more selective as foraging distance increases. The herbivory practiced by beavers is analogous to coppicing, an ancient forestry practice that involves cutting trees down close the ground to induce growth of basal sprouts, which are allowed to grow until suitable for harvest. Woodlands systematically coppiced can remain productive for centuries, just as beaver-cut willow thickets can remain healthy for centuries. Beaver activities along a valley bottom can also benefit willows by creating bare, moist soil for seed germination. The blockages created by beaver dams help to increase late summer and autumn soil moisture levels, which helps willow seedlings survive. And, by cutting willows stems in a manner that leads to asexual reproduction of willows, beaver cutting encourages continued reproduction of the willows. Reading the scientific papers describing these relations, I am tempted to regard the beavers as very skillful gardeners who manipulate the surroundings to indefinitely maintain their own food and building supplies. No wonder we call beavers ecosystem engineers. Biologists have another term for the relationship between beavers and willows, describing them as mutualists that can likely persist indefinitely within the same reach of a stream.

Once large numbers of elk take up residence along a stream, however, the sustainable adjustment falls apart. Elk browsing during summers is particularly detrimental to beavers. Just like human gardeners growing summer produce to can or freeze for winter, summer willow growth provides an important source of beaver food on stems cut during the autumn for a winter food cache. Summer elk browsing can also be more detrimental to willows than browsing during the plant's dormant season, and elk browse the same plants more frequently than do beavers. The frequency of beaver cutting of an individual willow is relatively low. The beaver has to wait several years while the willow regrows a stem large enough for the beaver to risk predation and expend energy in carrying the stem to a safe eating site, winter food cache, dam, or lodge. Beaver may also avoid juvenile willow sprouts because these contain a higher concentration of unpalatable chemical compounds. Whatever the cause, willows can rapidly recover stems cut by beavers, and the frequency of cutting is low. Consequently, the plants can recover lost mass and height and are more likely to reach sexual maturity and produce seed than are willows browsed by elk and moose. Baker and his colleagues concluded that willows can tolerate either complete, infrequent cutting by beaver or partial, frequent browsing by elk and moose, but not both.

The beaver population censuses and historical air photos tell the story. Beavers were once abundant in Moraine Park. Today there are none. Each year between 1968 and 1992, elk browsed, on average, 85% of the growth shoots on riverside willows. Between 1937 and 1996, the area covered by willows taller than 10 feet declined by 54% in Moraine Park and by 65% in Horseshoe Park. Today it is easy for park visitors to see the elk because there are few willows to hide them. There are no beavers to be seen.

Bruce Baker concluded that a lack of willows suitable as winter food for beavers can cause beaver populations to decline. This is a slippery slope for both beavers and willows, because the absence of beaver cutting creates a feedback mechanism that reduces willow resprouting and causes declines in beaver and willow populations.

Ecologists recognized that a long slide down this slippery slope can create a situation from which beavers and willows can only recover if something removes the elk and moose. This is not unique to Rocky Mountain National Park. After gray wolves were extirpated from Washington's Olympic National Park during the early 1900s, intensive elk browsing significantly reduced the growth of black cottonwood and bigleaf maple seedlings and sprouts. As river environments within the park became denuded, similar changes did not appear beyond the park boundaries along rivers protected from elk grazing. Channels inside the park became wide and shallow, braiding into secondary channels among eroding banks. Ecologists Bob Beschta and William Ripple described this scenario as an example of a truncated trophic cascade—a technical name for removing the top predators in a food web, with the consequence that populations of prey species grow rapidly and denude their own food sources.

Similarly, extinction of wolves in Yellowstone National Park allowed growing elk populations to outcompete beaver and heavily browse riverside willows. Riverscapes formerly occupied by mosaics of willow thickets, beaver ponds, and branching channels changed to drier grasslands with few willows and a single channel cut deeply into the sediments that had once accumulated behind beaver dams. Ecologists call these drier valley bottoms elk grasslands.

Elk grasslands and beaver meadows are examples of alternative states. As long as elk keep browsing the riparian zone, willows and other woody riparian shrubs and trees cannot regrow. In the absence of winter food and dam-building materials, beavers cannot survive in the valley bottom. Without beaver dams, high flows concentrate in a single channel, their energy widening and deepening that channel and limiting the spread of floodwaters across the floodplain. The valley bottom gradually grows drier. As the soil dries, burrowing rodents move in from the adjacent uplands, spreading via their feces the spores of fungi that spruce and fir need to draw nutrients from the soil. Upland conifers invade the valley bottom, but they provide limited beaver food. The elk grasslands remain a dry environment inhospitable to beavers.

As long as beavers are present, on the other hand, their dams block the channel and cause high flows to spread across the valley bottom. Water seeping into the soil

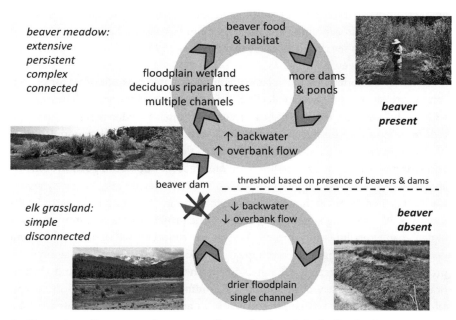

An illustration of beaver meadows versus elk grasslands and the feedbacks that occur to maintain each of these conditions. A color version of this figure is included in the insert section.

maintains the wet environment in which willows and other woody riparian plants thrive, thus ensuring a supply of food and building materials for the beaver. Beavers move about the valley bottom through time, creating diverse environments of new ponds, abandoned ponds gradually filling with sediments, small side channels, and the drier berms of old, abandoned dams. The beaver meadows persist.

Each of these alternative scenarios can remain stable for decades or, in the case of the beaver meadows, centuries or millennia, unless something external interferes. That something is most likely to be predators, either natural or human. Removal of predators and protection of elk in national parks has allowed the animals to create elk grasslands. Reintroduction of predators does not eliminate elk populations, but it does keep the herds sufficiently mobile to reduce browsing pressure along rivers and facilitate regeneration of willows.

Yellowstone provides the now-classic example in which wolf reintroduction indirectly created a dramatic change in river corridors, with willows gradually growing back into river-bottom thickets and beavers building dams and creating floodplain wetlands. Restoration of predators alone, however, was not sufficient to restore riparian zones along small streams in Yellowstone. Instead, the return of the willows depended both on limiting elk browsing and on restoring the water levels present before the removal of wolves.

Gray wolves (*Canis lupus*) were reintroduced to the northern portion of Yellowstone in 1995. Growth of the wolf population during 1995–2000 coincided with—but did not cause—a 70% decline in elk numbers. Restoration of tall willow communities requires that the plants exceed about six feet in height. Stems on tall willows exceed the reach of browsing elk and can therefore provide a reliable seed source for new willow plants as well as preventing complete consumption of existing willows during severe winters. Ten years of protecting willows from elk browsing was not enough to allow the willows to grow beyond the six-foot-height threshold for recovery unless the water level in the valley bottom was also raised by simulated (i.e., human-built) beaver dams. Where such dams were built, elk-browsed willows were nearly half again as tall and had 90% more mass than elk-browsed willows along portions of the stream without such dams. In other words, simply keeping the elk moving was not enough. The loss of beavers and their dams from the network of small streams in Yellowstone had lowered valley-bottom water tables to the point that willow recovery was hampered unless people stepped in and mimicked the effect of beaver dams, creating the willow growth and other conditions that would support the return of the real dam-builders.

Analogs for the dichotomy of beaver meadows and elk grasslands come from other rivers far too large for beavers to dam. Daniel Kroes and Cliff Hupp have shown how channelization of Maryland's Pocomoke River has also created something like an elk grassland. Channelization involves making a natural river more like an irrigation canal by widening, deepening, and straightening the river, as well as removing instream obstructions such as downed wood. Rivers around the world from small prairie and mountain headwater streams to the lower Mississippi River have been channelized for reasons including flood control and navigation. Just as when beaver dams are removed, channelization allows high flows to be contained within the channel, limiting connectivity between the channel and the floodplain and gradually drying the floodplain. Where channelization is combined with de-liberate drainage of wet valley bottoms for agriculture or other development, the rich, black muck of the floodplain is exposed to oxygen, releasing stored carbon as the soil dries and subsides. Burrowing organisms such as ants, termites, mice, and voles move in once the floodplain dries, further exposing the deeper sediments to oxygen and oxidation. Floodwaters, now confined to a single, straight channel, rush downstream faster and with more energy to erode the stream banks. The net effect is to completely transform the valley bottom via an activity as seemingly limited as straightening or deepening the main channel.

The key of course is that a stream or river does not simply flow passively through a landscape, but rather interacts closely with the surroundings, shaping the valley bottom and valley sides, and shaped by them. Alter one of these components and you are likely to see alterations in the entire system, as well as in upstream and downstream portions of the river network. This process of thorough transformation has been called river metamorphosis. There are many ways to force a river across some invisible threshold and trigger metamorphosis, including channelization and removing beavers.

Restoration along some of the world's big rivers—the Mississippi, the Missouri, the Illinois, the Danube, the Murray-Darling—now focuses on undoing some of the negative effects of channelization by reconnecting the river channel with its flood-plain. Such reconnection is not easy, simple, or inexpensive. Along smaller streams, beavers can do the work for us if we give them a little help. As Enos Mills wrote more than a century ago: "Beaver would help keep America beautiful" (p. 220).

Nature Green in Tooth and Paw

I knew that my attitudes had become highly beaver-centric the first time I saw an elk in the North St. Vrain beaver meadow and spontaneously thought: the enemy! Of course elk are not the enemy, but my immediate reaction to seeing the elk, about which I had read a great deal, gave me pause. My attitude toward moose has also changed gradually as I have come to find more and more signs of moose in aban-doned beaver meadows on the western side of the park. Visiting a historical ranch site along the Upper Colorado River in this part of the park, I found it hard to get beyond focusing on the heavily browsed, stunted willows that dominated views of the riverscape from the trail.

A beaver can build a dam in the most surprising places—steep, narrow valley segments from near timberline down to the lowest elevations in the national park.

I have found abandoned, breached dams throughout the drainages of the park. The dams are likely abandoned and breached because the sites are too small to sustain an entire colony for a prolonged period of time: the narrow valley bottom supports only a few willows or aspens and has room for only a small pond. I imagine that yearling beavers emigrating from larger beaver meadows during periods of relatively high beaver population density constructed many of these dams. The valley segments sufficiently broad to support extensive stands of willow and other riparian deciduous trees, and therefore to support multiple beaver colonies for periods of decades to millennia, are much more limited in number. Rocky Mountain National Park does include many of these valley segments, primarily along the eastern and western boundaries of the park where the Pleistocene glaciers reached their lowest elevations and left moraines that helped to create a valley setting suitable for beaver meadows. The dark, organic-rich soils and breached beaver dams of these valleys record their former existence as beaver meadows: the Upper Poudre River, Upper Colorado River, and, coming south along the eastern side of the national park, the North Fork Big Thompson River, Cow Creek, Black Canyon Creek, the Fall River, the Big Thompson River, Glacier Creek, and North St. Vrain Creek. Of these, only North St. Vrain Creek still has multiple beaver colonies.

Broad valley segments occur beyond the park boundaries, as well, but these sites have largely been preempted for human use: the city of Estes Park sprawls across what were once extensive beaver meadows, and the water ponded behind Buttonrock Dam submerges another former beaver meadow along North St. Vrain Creek near the mountain front. A few beavers remain of the once-abundant populations that attracted 19th-century fur trappers to the large, high-elevation valleys of North, Middle, and South Parks, but much of the land in these parks is now urbanized or in private ranches and the human landowners may or may not be receptive to beaver activity.

This is why I consider it appropriate to intervene in natural processes occurring within Rocky Mountain National Park. The ideal of a national park protecting naturally functioning ecosystems free from direct human manipulation is intellectually attractive, but the current conditions of the ecosystems within Rocky Mountain National Park reflect more than a century of human manipulation, including fire suppression, extinction of wolves and grizzly bears, and reintroduction of elk and moose. Unless natural resources managers actively intervene, beavers may disappear completely from the park.

Active intervention can take more than one form. The fenced grazing exclosures in many former beaver meadows on the eastern side of the park are part of a long-term plan to restore woody riparian vegetation in these meadows. Controlled hunting of elk within the park and use of contraceptives are currently being employed to maintain elk herds at desired levels of 600 to 800 animals. Whether the exclosures and elk population control will be sufficient to allow beavers to return to some of the former beaver meadows remains to be seen. More active intervention

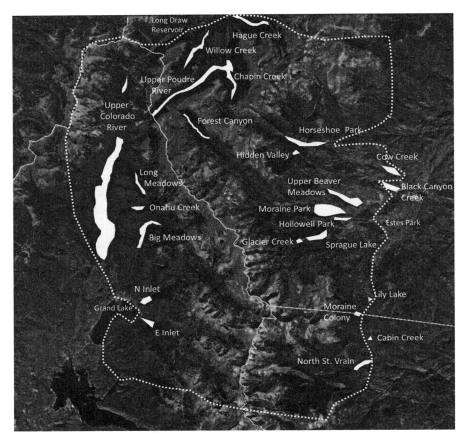

Schematic map of sites historically (yellow) and currently (green) occupied by beavers
in the vicinity of Rocky Mountain National Park (approximate park boundaries in dotted
white line). A color version of this figure is included in the insert section.

in the form of human-built temporary wooden dams to raise riparian water tables
or reintroduction of beavers can also be used, and the Park Service is now con-
sidering experimenting with human-built structures as a first step to eventually
reintroducing beavers. Most of the former beaver meadows in Rocky Mountain
National Park are now elk grasslands. Like Ebenezer Scrooge visited by the ghosts
of Christmas past, present, and future, the broad valleys of the park could go either
way—maintain their present status or cross a threshold and return to the beaver
meadows of centuries past.

I know which alternate reality I'm hoping for.

October

Of Beavers and Humans

By mid-October, the first snow has fallen on the beaver meadow. There is no sign of snow when I visit a few days later, but the air feels chill in the shadows and a cool breeze leavens the sunshine's warmth. Mostly, the beaver meadow seems a golden place. Many of the willow, aspen, and birch leaves have already fallen, but enough remain to create a glowing ménage of yellow, gold, palest orange, and tan. Each leaf refracts and filters the light so that it comes from every direction rather than only from above. Aspens on the north-facing valley slope stand bare and pale gray. Those on the south facing slope form bursts of gold among the dark green conifers.

The beaver meadow remains lively with activity. Dance flies move upward and downward in a column of air backlit by sunshine, their delicate bodies shimmering in the low-angle light. A little black stonefly lands on the back of my hand. I resist the urge, bred by summer mosquitoes, to reflexively slap it away. As I cross smaller side channels, brook trout dart away from the warm shallows where they have been resting. The narrow band of white on each dorsal fin flashes as the fish moves swiftly toward deeper water. When one small trout gets momentarily stuck between two exposed cobbles, I cup its slender, wriggling body between my hands and help it along. Windrows of fallen leaves form swirling patterns on the water surface and streambed. Filamentous algae grow in thick green strands along the side channels, where lower water exposes wide bands of mud along the channel edges. The mud bands record the comings and goings along the channel: precise imprints of raccoon feet and deer hooves and blurrier outlines left by moose. Moose beds mat down the tall grasses scattered among the willow thickets.

As usual, the beavers themselves elude me, but I see fresh mud and neatly peeled white branches with gnawed ends on some of the dams. Lower water in the beaver pond exposes an entrance hole in the side of the lodge. A fresh mudslide down the lodge suggests the occupants have been winterizing their home. The beavers should

The beaver meadow in mid-October, with snow on the high peaks and the willows and grasses in their autumn colors. A color version of this figure is included in the insert section.

be more active now as they repair dams and lodges and gather a food cache in preparation for winter. One of my colleagues was here during the snowfall earlier in the month and saw a beaver walking up the steep, rocky trail toward the upper part of the river basin. This surprises me for multiple reasons. I know that two-year-olds leave the home colony to found their own lodges, but typically they do this in spring or early summer. In addition, such a long trek away from water seems very risky for a beaver. And it seems odd that the animal would be following a trail used by humans, although few human hikers were out on that snowy day in autumn. I was slightly inclined to doubt my colleague's story until, the following summer, I saw freshly cut aspen trees and beaver trails along Ouzel Creek well upstream from Ouzel Falls. At least one intrepid beaver clearly made the climb. I hope that it prospers in its new, high-elevation home.

For now, I sit down at the edge of the pond in the main beaver meadow along North St. Vrain Creek to see what might materialize. Dragonflies alight on dried grass stems colored a paler reflection of the golden aspen leaves. Fresh snow on the high peaks up the valley reflects in the dark water of the beaver pond. Long shadows stretch northward from the valley wall, where the pond has an edge of thin ice.

The occupants of the nearby trees are vocal. A squirrel chitters at me for being too close, then returns to systematically stripping a spruce cone. A thick, spongy pile of chewed cones at the base of the tree is pitted with little holes where the squirrel is burying its winter supplies. The breeze carries calls of ravens and Steller's jays, but they are drowned out by a noisy Clark's nutcracker that darts back and forth among the trees across the pond. I hear a sound that I don't recognize in the understory below the nutcracker, a sound that I decide to call a quorkle. The quorkle becomes a sort of moof—and then a huge bull moose walks slowly out of the forest. I am sitting down and remain very still. The moose eyes me for a few moments, then moves to the edge of the pond to browse on a willow. A broad rack of antlers with tines polished palest gray spreads up and out from his head. It seems as though the weight of them would topple the moose when he bends down to drink from the pond, but the animal doesn't even spread his front legs. A spider crawls steadily up my thigh and I finally move to brush it off. The moose gives me another stare before he continues browsing the willow. I get up slowly and move quietly away, unsure of his temperament during the rutting season.

Bull moose at the beaver pond. A color version of this figure is included in the insert section.

Lessons Not Yet Learned

Meanwhile, the demand for water to irrigate crops and suburban lawns at lower elevations has largely stopped for the year. As the population density of Colorado rapidly grows, suburbs and cities steadily replace farm fields. This reduces per capita water consumption but creates immense pressure for urban water planners to ensure viable water supplies for continued growth. This in turn leads to more diversions and pipelines and more calls for reservoirs—following the same patterns used in Colorado for more than a century—but not yet to more calls for sustaining headwater regions by protecting and restoring beaver meadows. We seem to so readily forget or ignore the lessons learned from past beaver reintroductions.

Eric Collier's 1959 book *Three Against the Wilderness* provides a nice account of one such reintroduction. Collier settled with his wife and son in the woods of British Columbia, Canada, during the 1920s. His story of their stay in the region includes remarkable descriptions of the effects of beaver reintroduction. Collier was keenly aware of the ability of beavers to maintain a wet landscape even during episodes of drought or wildfire. Viewing an abandoned beaver meadow after a wildfire, Collier wrote of how the scene would have differed if the beavers had remained present to maintain the dam and associated wetlands and concluded "with the beavers gone, all hope seemed to have gone from the land too" (p. 16).

Collier made a living trapping and writing, but cattle ranchers who needed both forage and water to stay in business surrounded him. He described his first sight of the valley where he settled, when flow in the creek barely supported a first crop of alfalfa that provided winter feed for cattle. Collier reintroduced beaver to the creek, celebrating as the beavers built dams and began to pond water until, after a few years, the transformed valley bottom held sufficient water even in the driest summer to support a second crop of alfalfa in lower portions of the valley.

Collier was practicing the approach advocated by Archibald Belaney, an Englishman who lived in northern Canada for many years. Better known as Grey Owl, Belaney headed beaver conservation programs in two of Canada's national parks and observed beavers closely. His work and his books inspired beaver reintroduction and conservation programs that became popular across North America following the widespread drought of the 1930s.

The welcoming attitude toward beavers of Eric Collier and Archibald Belaney is still not universally shared. Although people are increasingly aware of the beneficial effects of water engineering by beavers and the "cuteness" factor of the animals wins them fans, removal of beavers and their dams remains common and widespread. Enos Mills wrote a dramatic account of one such removal from Lily Lake, a few miles to the north of the beaver meadow on North St. Vrain Creek.

Visiting Estes Park in 1872, the wealthy British Earl of Dunraven was taken by the beauty and hunting opportunities of the region. Using dummy purchasers, he quietly set about buying as much of the land in the area as he could, with the intent of establishing a private, European-style hunting preserve. When local settlers realized what the Earl was up to, they blocked further land sales and limited the extent of his domain, but by 1877 the Earl was sufficiently well established to build a large hotel for other visitors. The Earl directed some of his employees to dig a canal to convey the waters of Lily Lake down to his fish-rearing ponds in Estes Park. A colony of beavers occupied Lily Lake, and Enos Mills recounted repeated back-and-forths as the beavers dammed the canal, the men removed the beaver dam, and the beavers rebuilt the dam. Eventually the men started trapping and shooting the beavers, although other beavers from the colony continued to rebuild the dam. The contest continued until the men killed six beavers. The other beavers abandoned the site, an event that Mills described with pathos:

> One night . . . an unwilling beaver emigrant party climbed silently out of the uncovered entrance of their house and made their way quietly, slowly, beneath the stars, across the mountain, descending thence to Wind River, where they founded a new colony. (p. 182)

I have watched a modern version of this drama play out along Fish Creek, which drains from Lily Lake into Estes Park. In 2011, beavers built a dam and lodge along the creek within the aspen grove. I knew that beavers could be tolerant of traffic and human presence, and during my regular summer commutes between my home in Fort Collins and the North St. Vrain watershed, I watched with interest as the dammed area expanded.

Before the 2013 flood, I gave a talk at the Estes Park Public Library about historical changes in rivers of the region. I got into an interesting conversation with residents of the Fish Creek neighborhood after the talk and learned that the activities of the beavers had triggered an ongoing debate among the human residents. People on one side of the debate were delighted at the beavers in their midst but were concerned that sale of the private property on which the beavers lived could jeopardize the existence of the colony. Those on the other side of the debate did not object to the beavers per se but wanted to either remove the beavers or somehow control their activity in order to preserve the lovely riverside aspen grove that the beavers were steadily chewing their way through.

The next year, I noticed that some of the aspens had protective wire-mesh fencing around the base of the trunk to keep beavers from felling them. Otherwise, the colony remained. Then came the flood of September 2013. The flood caused extensive erosion of stream banks along the length of Fish Creek, as well as removing sections of the road, various pipelines buried under the streambed or in the banks, and some of the houses built too close to the creek. The flood also swept away the beaver dam and lodge, but some of the beavers survived. As I described in the March chapter, at least some of the damage to the road and infrastructure along Fish

Before the September 2013 flood: a for-sale sign in front of the beaver pond and lodge along Fish Creek. A color version of this figure is included in the insert section.

Creek might not have occurred if an active beaver meadow had helped to slow the downstream movement of the flood peak.

Despite the beneficial flood control provided by their work, even the beavers of the North St. Vrain colony are not fully protected from human predation. A paved road and a culvert at the eastern boundary of the national park artificially divide the beaver meadow, which continues downstream for another 2,000 feet to the Pleistocene glacial moraine. This lower portion belongs to the owners of the Wild Basin Lodge, who rent their facility for events such as weddings. Annoyed at the felling of aspens around the lodge, the owners remove the beavers. Relocation is not quite as straightforward, however, as in Enos Mills's day. Colorado Parks and Wildlife must be involved, and there are few wild places to which beavers can be relocated or to which they can relocate themselves. Under these circumstances, the beavers are killed rather than relocated.

This creates a situation that ecologists refer to as sources and sinks. High-quality habitat that allows populations of a particular species to increase is a source. A sink is lower-quality habitat that would not be able to support a population on its own, although the population in a sink can persist if individuals continually move there from a source area. The portion of the North St. Vrain beaver meadow within Rocky Mountain National Park is a source sending out pioneering yearlings each spring.

Those unfortunate enough to move just downstream into the tempting habitat beyond the park boundary enter a sink not because of habitat quality but because of hunting pressure. Fortunately, some of the beavers are also expanding upstream to sites with old, abandoned dams but without beaver activity during the past two decades. Perhaps the North St. Vrain colony will reclaim old territory.

After the September 2013 flood, beavers returned to the Fish Creek site to start rebuilding their dam. This time, however, the beavers faced a force much more implacable and persistent than floodwaters. The human property owners, determined to develop the site for housing, removed the animals.

The Fish Creek residents and the Wild Basin lodge owners worried about preserving their aspen grove represent one of the reasons people commonly remove beavers and their dams. Another is the flooding created by beaver ponds, which can grow large enough to overtop roads and property adjacent to creeks. Clever human engineers have developed solutions to these problems that are less draconian than killing the beavers. Small fences tightly surrounding the base of each tree can protect riverside trees. Water level in a beaver pond can be controlled with a leveler, a flexible pipe that runs through a beaver dam and drains the pond quietly so that the water level cannot rise above a certain elevation and the sound of flowing water does not trigger additional dam-building by the beavers.

A third concern with beavers is that the animals will either build a dam inside a culvert or that wood dislodged from the beaver dam will float downstream and obstruct the culvert. To prevent this, there are beaver deceivers and beaver stoppers. A beaver deceiver is a fence-like exclusion cage of posts and welded wire mesh, typically about 30 to 130 feet long, that is placed before the upstream opening of a culvert to keep beavers from blocking it. A beaver stopper is a wire-mesh cylinder inserted into a culvert, with a second, larger wire-mesh cylinder over it.

Yet another approach is to use vertical culverts that function like elevated bathtub drains. Nothing happens as long as the water level is below the upper end of the culvert. When the water rises high enough to overtop the culvert, water drains down and through a connected horizontal pipe beneath the road. Pierre Bolduc, a modern-day Eric Collier who lives in Alberta, persuaded his county to use this style of culverts near his property. The culvert prevented a flood even when the beavers' 300-foot-long dam breached in May 2016 and drained a large amount of water.

Technology exists that allows people to continue to build their infrastructure near active beaver colonies and the technology is not particularly expensive or complicated. The primary limitation seems to be human perception and desire. Describing beavers, Rachelle Haddock of Alberta's Miistakis Institute said, "I like to call them the wolves of the wetlands. People either love them or hate them."

A study of beaver populations along an urban–rural gradient in Massachusetts during 2001–2004 found that more beavers survive in suburban lands, where the primary dangers come from disease and accidents, than in rural areas, where death

An example of a structure that prevents beavers from damming the inlet of a culvert underneath a road. At this site in the Green Mountain National Forest of Vermont, a simple wire-mesh cage prevents floating wood or beaver dams from blocking the culvert.

results primarily from trapping and shooting. Strict guidelines on beaver relocation in many states unfortunately make it easier to just kill the beavers.

Something like a beaver exchange market is gradually developing, however, as individuals like Sherri Tippie of Colorado facilitate the movement of beavers from places where landowners want to remove beavers without killing them, to places where landowners would like to reestablish beavers. Tippie runs the organization Wildlife 2000. The PBS documentary "Leave It to Beavers" describes her work during more than 30 years of relocating beavers.

Beaver Leas

People and beavers have a long history. Hunter-gatherers have targeted beavers in Eurasia since the Ice Ages, which is part of the reason that only an estimated 1,200 beavers remained in Eurasia, divided among eight small populations, by the start of the 20th century. On the other hand, creation stories for many Native American tribes describe beavers helping the Great Spirit to shape the world, and beavers were commonly kept as pets around Native American villages.

In addition to hunting beavers, early Europeans also took advantage of the eco-system engineering of the beavers. Clearing the forests of northwestern Europe to create crop fields and grazing lands was a major challenge for famers prior to development of metal tools. In 1983, two British archeologists proposed that Stone Age hunters and farmers took advantage of beaver meadows as sites already cleared of trees. Working at the Somerset Levels site in England, J. M. Coles and B. J. Orme found that Mesolithic and Neolithic people used willows felled by beavers to build a shoreline platform for hunting. Beavers were present in Britain starting soon after the glaciers retreated. Numerous archeological sites suggest that Mesolithic people sought out these locations, which provided grazing for domestic animals and attracted wild herbivores that people hunted. Coles and Orme describe the prehistoric forest as being moth-holed with beaver-created clearings.

Hunted nearly to extinction by the early 17th century, European beavers are gradually coming back. Now that they are rare, new colonies make the news when they appear in Germany or Britain. However, contemporary populations of *Castor fiber* estimated at more than a million animals are likely much, much lower than those present a few centuries ago.

When Europeans began to explore North America during the 17th century, they came into contact with extensive, complex beaver meadows on a scale not seen in Europe for more than two centuries. The budding field of natural history gave rise to several descriptions of beaver colonies and engineering. Margot Francis traces the history of these descriptions, many of which used the beaver as a metaphor for human society, whether monarchical (Nicholas Denys wrote in 1672 of the "laborious and disciplinable nature [of a beaver] and its obedience in work"), democratic (Baron de Lahontan in 1705 wrote of a beaver republic), or patriarchal (Bacqueville de la Potherie in 1722 wrote, "His house is so admirable that one can recognize in him the authority of an absolute master, the true character of the family Father, and the genius of a skilled architect"). Francis notes that narratives about beavers as intelligent, paternal, and hierarchically organized workers cover several centuries and constitute a collective image. The idea of a beaver as the "family Father" makes me smile, given contemporary understanding of the role of the adult female in overseeing lodge- and dam-building.

Appreciating Beavers

Today, beavers definitely have a fan base in the United States. I have published more than 200 technical articles in scientific journals over the course of my career, but the few articles I have written about some aspect of beaver engineering of rivers and landscapes have gotten far more public attention than any of my other work. When my colleagues and I published a short article about the total thickness of beaver pond sediments as revealed using ground-penetrating radar, the work was picked up by science news outlets on the web, natural history websites, and even a religious website. I started getting emails from beaver admirers, and now I receive notice of some beaver-related video or other piece of news on a weekly basis. Just as 18th-century writers celebrated beavers for the industry and organization viewed as desirable in human society, so we of the 21st century celebrate beavers for ecosystem engineering and the ability to contribute to "saving the planet."

On the other hand, plenty of people regard beavers as nuisance rodents, to be controlled or manipulated in the same manner as any other rodent causing inconvenience to humans. The most striking example of this I have come across occurred during a Sierra Club trip into a remote provincial park in the Yukon Territory of Canada. During a 10-day backpacking trip that featured abundant rain, cold temperatures, and some very challenging hiking, everyone on the trip remained cheerful, delighted to be able to explore the rugged scenery of a place so little altered by humans. At supper, one of the trip leaders mentioned that once he returned home he needed to remove a beaver dam and colony that was "ruining" the creek on his property. I was appalled that someone I expected to have more environmental literacy either was ignorant of the importance of beavers in maintaining the health of the creek or else was willing to ignore the role of the beavers simply to create a riverscape that he found more esthetically pleasing. The incident reminded me of Aldo Leopold's words in *A Sand County Almanac*: "Recreational development is a job not of building roads into lovely country, but of building receptivity into the still unlovely human mind" (p. 295).

One of those whose receptivity to beavers increased with time is Lewis Henry Morgan (1818–1881), author of *The American Beaver and His Works*. Morgan was an American scholar who invented the study of kinship as the analysis of family relationships and in so doing founded the discipline of anthropology in the United States. Animal behavior also fascinated Morgan, and his systematic studies on this topic culminated in his beaver book. He was an early proponent of the importance of intelligence, rather than simply instinct, in guiding animal behavior, and beavers provided his prime example. As traced by Gillian Feeley-Harnik, Morgan's beaver book grew out of a visit to northern Michigan during August and September 1860. Traveling with woodsmen of French and Ojibwa descent, Morgan studied beaver dams and ponds. He was particularly struck by the complexity and longevity of the beaver meadows he visited, and from these features he inferred not only intelligence but consciousness of self for beavers. Hope Ryden's careful observations of the beavers in the Lily Pond colony support Morgan's inferences. The Lily Pond beavers played, learned, expressed alarm and pleasure, and formed strong bonds with one another and with human observer Ryden. Reading Ryden's descriptions of the beavers' behavior, I found it difficult not to conclude that the animals have some consciousness of self.

Although scientists continue to debate whether non-human animals possess consciousness, no one debates the plethora of beneficial effects that beavers create within the environments in which they live. To paraphrase a sentence from Morgan's book, when a human stands for a moment and looks upon the work of beavers, a good deal more than a pond dammed by sticks is contained within the view. All of the diverse features within the beaver meadow together create a whole greater than the sum of its parts. Yet this is still only the surface view.

Hidden within this surface view is the storage of water in the meadow sediments that partly hold and then gradually release each spring's snowmelt and each summer thunderstorm's downpour, keeping the meadow verdant during dry seasons and times of drought. Hidden within the view are the tons of sand, silt, and clay accumulating above the bedrock and boulders that floor the beaver meadow, as well as the nitrogen, carbon, and phosphorus that cling to the sediment grains and the decaying leaves and twigs that keep the sediment a rich black color. A flowering orchid or the insistent call of a wood frog on a spring evening may hint at the biological richness within the meadow, but the abundance and diversity of the microbes that power so many of the chemical transformations or the insects that form the base of the aquatic and riparian food webs remain largely hidden from view. In the North St. Vrain beaver meadow, at least, the beavers themselves remain largely hidden from view, although leaving evidence of their activities across the meadow. And, hidden within the view of the meadow as it is at present, lies the history of the Northern Hemisphere since the Pleistocene glaciers retreated more than 10,000 years ago: a history of plants recolonizing hillslopes and river valleys, providing the food and building materials needed to support beaver colonies, and of beavers laying sinuous lines of wood and earth across the landscape, building the support framework for the wetlands once abundant across Eurasia and North America. And then a history of the fur desert first created by sustained, systematic trapping and then maintained by deforestation, agriculture, land drainage, channelization, and urbanization and—ultimately—more than seven billion people and their inexhaustible appetites for more land, more resources, more, more.

I do not know if beavers have consciousness of self, nor do I ultimately care. They have an inherent right to exist on Earth, and other creatures, including humans, are much richer for the presence of beavers.

November

Beavers to the Rescue

By late November, snow covers much of the beaver meadow. I visit on a sunny day well above freezing, but the low-angle light comes with long, long shadows. The meadow is noisy with continuously rushing wind that keeps the bare willow branches swaying and sculpts the snow on the lee side of plants into streamlined mounds. Individual grass stems have traced downwind crescents on the snow surface. Tracks of wind, tracks of animals: the activities of the meadow are once again made visible in the footprints of moose, hare, squirrel, coyote, and birds. The snow is mushy in the warmth and many of the tracks are blurred, but I also cross fresh, sharply defined traces left by four little leaping paws, with just the brush from a long, slender tail behind them. The prints are so delicate that they barely indent the snow, but clearly a mouse was stirring here recently. The fragile tracery of tiny claws in the snow seems vulnerable, but I know the animal is probably better adapted to the cold than I am.

The main channel of the creek remains open, the water golden brown between white banks bulbed with ice along the edges. The creek flows quietly, the sound of moving water submerged beneath the wind. The larger side channels also remain open and green with filamentous algae, but I break through the snow-covered thin ice on the smallest side channels. The off-channel ponds are frozen more solidly. Mats of dried algae quiver in the wind on one newly drained pond. Downwind, the snow is dirty with silt blown from the exposed bed. A layer of sticks, sand, and muck floors the pond with a woody carpet created by the beavers.

The main beaver lodge is freshly plastered with mud and sticks, but the ice on the surrounding pond remains unbroken and the snow is trackless. Away from the pond, snow into which I sink to mid-calf obscures the details of the ground. The upright stems of willows and aspen trunks dominate the foreground. Between these vertical lines the beavers have laid down horizontals by felling several aspen trees. The snow is untracked around the stumps, although the wood appears freshly gnawed. I contemplate these once-solid pillars of wood, several inches in diameter, and wonder how long it took a beaver to gnaw through them. Some of the newly downed trees also have fresh elk-tooth scrapes.

Despite the brightness of sun reflecting off snow, the winter colors of the meadow are subtle. Weathered gray bark covers the older willow stems, although the branch tips remain greenish-gold or pale orange. Beyond the meadow, the horizons are blue and white where the snowy peaks meet the sky.

There is no sight or sound of birds in the meadow, only the creaking of a stiff birch trunk in the wind. The wind is less noticeable in the pine forest bordering the meadow, and I come across a flock of chickadees moving among the pine boughs. Freshly stripped stems of pine cones at the base of a large tree record the recent presence of the squirrels that are usually so evident here.

Linear mounds of soil displaced by pocket gophers snake across the south-facing ground where the snow has melted. The gophers create a network of tunnels at the base of the snowpack. Where the animals burrow into the soil, meltwater running along the surface beneath the snowpacks displaces soil into the tunnels. Complete melting of the snow reveals these sinuous tubes of sediment lying on intact ground.

I enjoy reading these signs left by the varied inhabitants of the beaver meadow. After nearly a year of visiting the meadow, I have yet to catch a clear glimpse of the

Soil displaced by pocket gophers lies revealed on a patch of bare ground beneath the pine groves bordering the beaver meadow. The pencil provides a sense of scale. A color version of this figure is included in the insert section.

animals that I'm beginning to think of as the phantom beavers, but I take pleasure in knowing that they are leading their lives here.

Leave It to Beavers

Increasingly, people are deliberately reintroducing beaver. Although the intent may be not so much to enhance the beauty of the landscape through new beaver meadows as to store sediment, enhance stream base flows, or increase riparian habitat, the beauty comes along as an added benefit. Beavers have been invited to restore beaver-impoverished streams at least since work by the Soil Conservation Service (SCS) during the 1930s and Eric Collier's work along his neighboring creek during the early 1940s. What is particularly interesting is that people undertaking soil conservation and river restoration seem to rediscover the abilities and effectiveness of beavers every few decades. During the second half of the 1800s and first decades of the 1900s, writers such as Morgan (1868), Mills (1913), and Dugmore (1914) emphasized how much beavers had shaped lowland landscapes in North America and how they could help to conserve soil and water. These insights seem to have been forgotten for a couple of decades until the "Dirty Thirties," when resource managers were on a mission to reduce soil erosion. Paul Scheffer's description of the SCS's work along Mission Creek in Washington provides an example.

The SCS established a camp along the creek for "flood-control work" in response to damaging floods attributed to overgrazing and logging in the watershed. Scheffer described how the 1935 discovery of a few beavers in the upper waters of Mission Creek "doing effective work in the control of soil and water losses" inspired the introduction of 12 more beavers to the area by the Washington State Game Department. When the SCS surveyed the site in 1937, they found 60 dams along five miles of the creek. Most of the ponds were filled with sediment, and the creekbed had been widened from a few feet to more than 100 feet. The beavers had been so effective in limiting erosion that the SCS promptly began transplanting them to other sites, including Ahtanum Creek near Yakima, Washington. I can just imagine the beavers shaking their heads as they had to build one dam after another to hold back the soil that the humans had carelessly allowed to erode from the uplands.

James Grasse described how beaver "planting" on Wyoming's Coal Creek in 1941 resulted, by 1951, in a pond covering 25 acres that was heavily used by waterfowl, muskrats, mink, moose, and trout, among other creatures. Grasse's 1951 article in the *Journal of Forestry* extols the role of beavers in watershed protection, water conservation, and habitat improvement, which Grasse felt entitled the beavers to "a great deal of consideration."

Then there was another lull in appreciation for beavers as agents of erosion control until professional land managers once again began to use beavers in river and rangeland restoration during the early 1980s. Resource specialists from the U.S.

Bureau of Land Management (BLM), for example, worked with private landowners along Sage Creek and Currant Creek, both in Wyoming, to reintroduce beavers. Each creek had cut down below the surrounding, sage-covered terrain. The willows that once grew along the creeks were gone and continued erosion of the stream banks kept the flowing water turbid with silt and clay. The BLM supplied the beaver and some cut aspen logs for dams and food. Sage Creek, which the landowners reported could be easily crossed on foot or horseback a few decades earlier, had cut a 60-foot-deep trench that was more than 80 feet wide in places. Spring snowmelt flows were capable of removing beaver dams here, but two fish and wildlife biologists decided to try giving the beavers some novel dam-building supplies. The biologists dropped off a load of old truck tires, the adaptive beavers immediately incorporated them into new dams, and the dams held. Within three years, sediment loads on the creeks declined by 90%. Willows returned to the creek banks and helped to stabilize them against further erosion. The water table under the creeks rose enough to support aquatic plants, creating marsh and wet meadow habitat that in turn attracted a variety of animals.

Steven Albert and Timothy Trimble wrote of another success story in New Mexico. Beavers were once found on perennial rivers throughout the southwestern United States, including the Zuni River near the Pueblo of Zuni in New Mexico. In 1985, resource managers began to selectively reintroduce beavers to degraded streams on the Zuni reservation. The beavers were commonly able to complete a dam within one to two weeks, and the dam immediately reduced the velocity of stream flow, increased sediment deposition, and raised the bed of a stream that had typically cut down into a steep-walled trench. As the beavers expanded their activities, their dams ponded water over progressively larger areas. Riparian vegetation came back and songbirds, fish, amphibians, deer, and elk, among other animals, noticeably increased in numbers within the river corridor. Albert and Trimble wrote a how-to manual, noting that it is best to move beavers during early spring, before litters are born, or late summer, when kits are grown and relatively independent. The reintroductions were most successful when three or more beavers were released into the same area. Albert and Trimble described how beavers can transform a narrow, ephemeral channel into "a year-round wetland that now supports willow flycatchers, snipe, an occasional yellow-billed cuckoo, and dozens of other birds" (p. 90). The beavers succeeded even when released into arroyos with barely enough water to keep a few small ponds filled year-round, so permanent flow is not a necessity for beaver survival.

The momentum created by these success stories carried over into the early 1990s in publications such as Rich Olson and Wayne Hubert's 1994 booklet through the University of Wyoming's Cooperative Extension Service, which described beavers as water resources and river habitat managers. Other than this, however, there was another lull of about a decade in which little was written about using beavers to actively engineer more ecologically diverse and stable river corridors.

A new round of awareness and appreciation of beavers seems to have begun recently with papers such as the 2014 review article by Michael Pollock and colleagues in the journal *BioScience*. During the two preceding decades, river restoration grew to a multibillion-dollar industry in the United States, Europe, and Australia. Effective restoration of rivers requires that river plants and animals interact with physical features such as stream flow and channel shape, which people manipulate as part of restoration. The *BioScience* paper has a nice figure illustrating these interactions for the scenario of beavers released into a stream that has cut a deep, narrow channel down which high flows rush energetically. If the beavers build a dam in such a channel, high flows are likely to blow out the dam or erode around its sides. This erosion, however, widens the channel and allows at least a narrow floodplain to develop along the active channel, although still below the surrounding land. Because the channel is wider, high flows move downstream a little more slowly, allowing the beavers to build wider and more stable dams. The ponds above these dams rapidly fill with sediment and are abandoned, but the sediment provides good establishment sites for plants, which then help to stabilize the sediment while the beavers build new dams. This sequence is repeated through time until beaver dams create a thick layer of sediment into which water infiltrates, raising the water table underneath the channel and floodplain. The river corridor becomes physically and ecologically complex, with dams, ponds, meadows and willow thickets on filled ponds, logjams, and secondary channels.

Case studies documenting exactly this progression through time are now becoming more common, such as one by Glenn Johnson and Charles van Riper on the Upper San Pedro River in Arizona. The San Pedro originates in northern Mexico and flows northward into Arizona and the drainage of the Colorado River. Extensive riparian forests of cottonwood (*Populus fremontii*) and Goodding's willow (*Salix gooddingii*) supported breeding bird communities noted for being among the most dense and diverse in North America, as well as abundant beavers that maintained floodplain wetlands. Beavers were once so common along the San Pedro River that fur trapper James Ohio Pattie named it the Beaver River in 1825. The name did not stick, and neither did the beavers, which were trapped to extinction in the region by the early 1900s. A predictable sequence followed. As valley bottoms became drier, channels cut down and widened into steep-sided arroyos. Stream flow assumed more of a boom-and-bust character of flash floods alternating with dry periods and very low flow. Finally, federal and state wildlife managers reintroduced beaver to the river during 1999–2002. By 2005 at least 40 beavers occupied 13 different sites. Bird censuses during 2005–2006 documented greater abundance and diversity of a wide range of birds, including several species of conservation concern, in areas colonized by beavers.

Can many small dams built along a stream by people create effects similar to those of beaver dams? Typically not. People commonly build even small dams designed to impound water for livestock in a straight line that completely crosses the valley

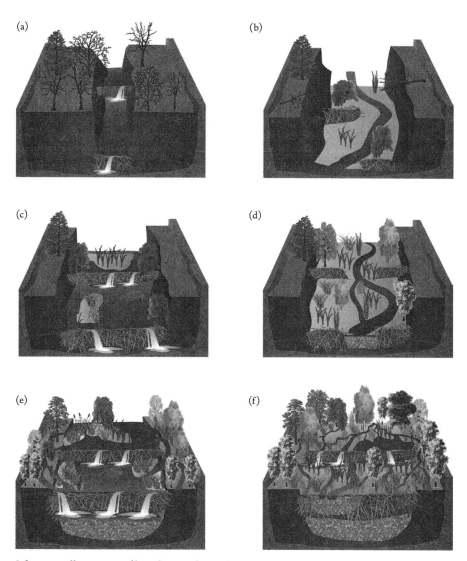

Schematic illustration of how beaver dams alter streams that have cut down into valley sediments. (a) Beaver dams in narrow streams are more likely to fail during high stream flows. (b) This results in erosion around the dam helps to widen the incised valley, allowing a narrow floodplain to form. (c) The widened valley results in less flow energy during floods, which allows beavers to build wider, more stable dams. (d) Streams that are incised commonly transport substantial sediment. Beaver ponds fill rapidly with sediment and the beavers move to another site along the channel, but the pond fill provides germination sites for vegetation. (e) This process repeats itself until the beaver dams raise the water table enough to reconnect the stream to its former floodplain. (f) Eventually, vegetation and sediment fill the ponds and the stream ecosystem becomes complex and resilient. (From Pollock et al., 2014, Figure 4.) A color version of this figure is included in the insert section.

bottom. The dam may be designed so that water can flow over the top, but it is not designed to be flanked by a secondary channel developing around the side. A traditionally designed, human-built dam ponds water behind it, creating the diversity of flowing and standing water within the river bottom, but secondary channels are not allowed to form, new dams are not built elsewhere at regular intervals while older dammed ponds fill with sediment, and the mosaic of habitats present in a beaver meadow does not develop around human-built dams. Recognizing these limitations, people are now deliberately mimicking the effects of beaver activities by building beaver-dam analogs.

Beavers to the Rescue

Federal agencies (the U.S. Fish and Wildlife Service, National Oceanic and Atmospheric Administration, and U.S. Forest Service) along with Portland State University and the North Pacific Landscape Conservation Cooperative released *The Beaver Restoration Guidebook* in 2015. Beaver restoration now has its own acronym, a sure sign of having "arrived" with the federal government. BDAs are beaver-dam analogs, the human-built wooden structures that I imagine beavers snickering at or scorning. In situations where beavers may not be able to survive initially because of limited food, BDAs can be used to jumpstart the process of beaver reestablishment. These temporary dams are designed to raise the riparian water table enough to support willows and other woody plants that can feed beavers. Some restoration efforts even create multiple, sequential BDAs along a stream, analogous to the stepped dams that beavers might build. BDAs have been successful in raising the riparian water table, increasing connectivity between the channel and floodplain, and luring the real ecosystem engineers back to a river segment in several restoration projects in Washington and Oregon. Use of BDAs along Bridge Creek, a tributary of the John Day River in central Oregon, has resulted in significant increases in production and survival of migratory steelhead trout (*Oncorhynchus mykiss*) as the quantity and complexity of habitat for the fish increased.

Bridge Creek hosts the largest number of BDAs. Michael Pollock and Utah State University ecologist Nick Bouwes installed 76 BDAs on two miles of the creek in 2009 and then added 45 more between 2010 and 2012. This spatial density is more like the closely spaced beaver dams historically present along many channels. Local beaver gave the effort their stamp—pawprint?—of approval by adding on to nearly 60 of the BDAs and building 115 new dams along the creek. The combined efforts of humans and beavers substantially expanded the extent of secondary channels and submerged areas, resulting in a tripling of steelhead numbers relative to a nearby unaltered stream, as well as greater likelihood of survival for the young steelhead from Bridge Creek.

Two examples of beaver dam analogs installed along Fish Creek in Estes Park, Colorado, after a large flood in September 2013.

The beaver restoration guidebook includes a beaver-dam viability matrix designed to help people evaluate whether a beaver dam is likely to persist over at least two seasons, the period required for a mating beaver pair to successfully rear their young. The matrix is based on physical factors such as valley form and characteristics of river flow and on adjacent land use and the presence of infrastructure such as bridges and culverts. Because beavers typically prefer relatively low-gradient stream segments with wide valley bottoms, their habitat overlaps heavily with agricultural and urban lands. This overlap can lead to conflicts when beaver ponds submerge roads and property or fell riverside trees that people would prefer to see alive and upright, or when the sound of water running into a culvert or an irrigation ditch triggers that irresistible urge and the beaver dams the flowing water. As explained earlier, each of these potential problems can be avoided or minimized with relatively simple precautions such as drainpipes in beaver ponds, wire cages to protect culverts, and wire fences around the base of each tree, but this requires planning and good will on the part of people. As my colleague Heidi Perryman of the Worth A Dam organization expressed it, "When you have aggravation, smart cities use education, mitigation, and relocation, before using termination." Organizations such as Worth A Dam are working diligently on education and outreach. The existence of how-to guidebooks and systematic evaluations gives me hope that this time around our efforts to work with beavers, or at least to get out of the way and let them work on their own, will be longer lasting.

The overlap between preferred beaver habitat and preferred human habitat can also be used to reduce some of the negative effects of human presence, given that our landscape engineering benefits very few species beyond ourselves. Globally, human activities now move more sediment than all non-human processes—wind, water, ice—combined. Our activities have hijacked the global nitrogen cycle, so that nitrogen emissions to the atmosphere and nitrogen entering rivers and oceans result dominantly from human manipulations. We dominate the movement of water at and near Earth's surface, pumping groundwater at unsustainable rates from large aquifers and storing a volume of water in reservoirs behind large dams that is equivalent to a 700% increase in natural storage within river corridors.

Although we have derived numerous local benefits from manipulating fluxes of water, sediment, nitrogen, and other materials, the cumulative regional and global side effects are becoming increasingly devastating. Fine sediment in the form of sand, silt, and clay is the most widespread pollutant in rivers of the United States, not because the sediment is inherently bad but simply because our clearing of native vegetation, agriculture, and reconfiguration of topography have introduced so much extra sediment into rivers that it accumulates even in large river channels. Excess sediment creates turbidity, smothers bottom-dwelling organisms, covers fish spawning gravels, and eventually changes channel form in ways that limit habitat for native plants and animals. Beaver dams and ponds also foster sediment accumulation, but on smaller channels including secondary, floodplain channels. The ubiquitous contemporary presence of fine sediment in large rivers such as the Illinois is what is now limiting habitat for native plants and animals. Similarly, although nitrogen is a vital nutrient for most organisms, excess nitrogen entering streams causes algal blooms, reduced dissolved oxygen in stream water, and hazards to human health.

Beavers and their dams are not a panacea for all that ails our rivers, but they can be a start toward river rescue. Beaver dams and meadows along rural and urban streams can be particularly effective in reducing the changes associated with human land use, including enhanced base flows during dry periods, storage of excess sediment, and storage and uptake of excess nitrogen. Julia Lazar and her colleagues measured nitrate uptake by three beaver ponds in Rhode Island. Extrapolating from these measurements to differing scenarios of beaver-pond distribution across the landscape, the scientists estimated that beaver ponds are capable of removing anywhere from 4% to 45% of nitrates from rural watersheds of southern New England that have problems with high nitrate levels.

Beaver ponds can also positively modify water chemistry in regions with acid rain. Scientists working in the Appalachians found that the ability of stream water to neutralize acids from atmospheric pollution increases following passage of the water through a beaver dam.

The ability of beaver dams and meadows to sustain base flows during dry periods is likely to be increasingly important in regions such as much of the western United States. Climate models forecast warmer temperatures and changes in precipitation from snow to rainfall, which moves through a landscape more rapidly than snowmelt, for many parts of the Intermountain West. Where beaver dams are persistent, the wet meadows that the dams create can moderate floods, augment base flows, and generally increase the ability of river corridors to withstand some of the projected changes in climate.

People seeking such benefits from beaver activity can work to increase landowner acceptance of beavers and let the animals disperse to new areas on their own, or they can actively identify all potential beaver habitat within a region and undertake targeted reintroductions. Joe Wheaton and William Macfarlane of Utah State University have developed a model for assessing the location and suitability of beaver habitat across a watershed or a large region. The model uses data available online, including maps of channels, riparian vegetation, and topography, as well as equations for predicting base flow and peak flows that occur on average every two years. Wheaton and his team found excellent agreement between model predictions and observed beaver dams in a test involving 1,210 miles of rivers in Utah. The best part of the model is its name: the Beaver Restoration Assessment Tool, aka BRAT.

Beaver reintroduction is not as simple, however, as dropping off a beaver along a river and letting the animal get to work. In 1948, the Idaho Fish and Game Department undertook a beaver-delivery experiment, boxing the animals up in wooden crates, attaching parachutes to the crates, and dropping them over national forestlands in Idaho where wildlife managers wanted to promote fur trapping. The assumption was that the crates would either break apart on impact or the beavers would chew their way out. Sixty years later, the promotional movie created at the time of the flying beavers had a revival on the internet. The movie is a good deal less amusing to watch when you realize that many of the beavers dropped from the sky probably died: mortality among reintroduced beavers is typically 80% to 90%. With very careful planning and management, some scientists undertaking restoration have achieved 50% survival of beavers.

The Beaver Restoration Guidebook includes detailed instructions for reintroducing beaver, including a section titled "Sexing Beaver." Given the territorial nature of beavers and the need to have males and females present to establish a breeding colony, knowing the gender of the individuals being reintroduced is critical. Beavers, however, do not possess the obvious displays of gender difference that make identification easier in some animals—there are no differences in markings, horns, or antlers, not even any external differences in genitalia. In what I imagine must be a very unpleasant process for both humans and beavers, the most reliable way to "sex a beaver" is to restrain the animal, manipulate the cloacal area to manually press out anal gland secretions, and then assess the color, viscosity, and odor of the secretion to distinguish males and females. I cannot improve on the wording in the guidebook describing this "tricky, slippery, and somewhat messy process . . . that gets easier with practice":

> You and your assistant should wear gloves . . . Have your assistant cradle and restrain the beaver on its back, in the [beaver] bag. Beavers do not like to be on their back and will thrash around if you let them. Cradling, as with an infant, can be effective . . . force applied in the wrong area . . . may cause the beaver to excrete feces or gas. Obviously this procedure should be done with your face a reasonable distance from the cloaca, with your mouth shut. You may also want to wear safety glasses . . . some animals will be easier to work with than others. Be patient and gentle; the animal under your care needs to be protected. It may happen that the anal gland withdraws and you will need to start over multiple times. (Woodruff and Pollock, 2015, pp. 73–74)

After that indignity, reintroduced beavers are particularly deserving of survival in their new homes.

Despite the heavy mortality, beaver reintroduction is commonly successful when numerous beavers are reintroduced to an area. A 1930s effort in Washington, for example, released 76 beavers at 40 sites: 18 of the animals successfully established a colony. Enough of the flying beavers survived that the experiment was deemed successful (although it was not repeated).

The greater obstacle to successful beaver reintroduction is likely to be human attitudes. Regulatory agencies sometimes cannot decide how to deal with a BDA because the structure is neither natural nor entirely human-built once the beavers modify it. The byzantine water laws of the western United States, which strictly apportion water use by seniority, are sometimes used to prevent construction of BDAs, even though the structures are designed to slow rather than stop the down-stream passage of water. And lingering negative views of beavers can prejudice landowners against anything associated with beavers. Journalist Ben Goldfarb described how one BDA advocate, Chamna Gilmore of California's Scott River Watershed Council, started calling BDAs post-assisted wood structures. I have also seen BDAs described as velocity inhibitors.

Nonetheless, BDAs are coming into their own and people are recruiting beavers as river restoration partners in portions of Europe. The European species, *Castor fiber*, went from nearly extinct at the start of the 20th century, with only 1,200 animals spread among eight Eurasian preserves, to an estimated 430,000 by the end of the century. In 2001, Peter Rolauffs and colleagues described the long-term success of beaver reintroductions undertaken in parts of Germany during the early 1980s. By 2009, 23 European countries were participating in beaver restoration. Beaver reintroductions began in 1974 in Poland, and by 2016 Dorota Giriat and other scientists wrote of 25,000 beavers busily building dams and beaver meadows in the more rural portions of Poland.

We are not currently in a period notable for widespread recognition of the problem of soil erosion, analogous to public awareness during the 1930s Dust Bowl. The U.S. Environmental Protection Agency, however, continues to list fine sediment as the most widespread pollutant in rivers. Beavers have proven their adaptability to natural challenges such as deeply eroded channels and high sediment loads. They have also demonstrated a remarkable ability to live in proximity to roads and suburban or agricultural areas with high human presence. As the momentum of the latest wave of beaver appreciation builds among scientists, resource managers, and the public, I hope that this time around it will be strong and broad enough to create lasting change in the way we manage river corridors and landscapes, as well as beaver populations.

December

Saving the Dammed

At the nadir of the year, this is how morning comes to the beaver meadow. Just as the sun rises above the eastern horizon, a flush of pale rose lights the snow newly fallen on the highest peaks. The beaver meadow remains in shadow, silent but for the creek flowing quietly between its rims of ice. The air temperature is well below freezing and frost whitens the pine needles like a dark-haired person starting to go gray. Wisps and sheets of snow flag off the summits in the steady wind. Over the course of a few minutes, the summit snow warms from pale rose to faint orange and then a rich, warm gold that also lights the rock outcrops at lower elevations. The wind reaches the beaver meadow before the sunlight, coming in abrupt blasts that shake loose the little tufts of snow remaining on the pine boughs. The wind sends the snow crystals slaloming across the ice on the creek with a dry, skittering sound like that of blowing sand. Before long, the meadow is submerged in a continual rushing sound created by wind gusting through the pines up slope, along the valley walls. The lateral moraine to the south keeps the beaver meadow in shadow until 9:30 a.m.

Nothing is so slow as waiting for the warmth of sunlight on a cold winter morning. When the sunlight does reach the meadow, it brings out the colors of water, ice, grasses, and willows. Flowing portions of the creek change from gray to orange brown. The snow reflects the light in a painfully intense glare broken by the deep, long shadows that everything casts. With the sunlight comes a steady wind that blasts the crystalline snow onto my face like grit.

Not much snow has fallen yet, but North St. Vrain Creek is completely frozen in places and covered with snow. The ice records the movements of water, freezing the pulses and turbulence in ice ripples and ledges, motionless swirls and bands. It seems a miracle that any water still flows in this gray and white world of ice and snow. Low water reveals the topography of the creek bed. Cobble bars rise like small hills, dry beside the narrowed flow that remains. And yet the pathways of the water seem even more mysterious than in summer, as when I watch the flow disappear under a ledge of ice. Liquid water appears where I least expect it. Snowshoeing across hummocky terrain well away from the creek, I break through into a slushy depression, completing the annual cycle of wetting my feet at least once each month of the year in the beaver meadow.

The surface snow is relatively fresh and strongly sculpted by the wind, but I still find the trace of a deer. The meandering track of a coyote brings to mind the title of a lovely book by Edwin Way Teale, *Wandering Through Winter*. Many of the tracks are so recent that I can clearly see the imprint of the coyote's pads and toenails. The animal explored the region where the beavers felled a dozen large aspen trees two months ago along the northern edge of the meadow, a region I now call Fallen Timbers after a place in the Midwest. In more sheltered patches beneath the trees

The history of water movement as recorded in ice, December in the beaver meadow. A color version of this figure is included in the insert section.

I find the delicate tracks of a small animal whose tail dragged slightly in the snow as it jumped. Even the weight of a small body compacts the snow enough to make it more resistant to the wind. On one of the exposed cobble bars, the wind has winnowed away the loose snow from the tracks of a small animal, which now rise slightly above the surrounding surface like a delicate chain mesh laid down across the bar.

Just as the ice records the movements of water, the snow records the movements of wind. Each willow thicket and small bush has a leeward island of snow. The beaver dam is covered in miniature ridges and swales that bear no relation to the underlying topography, as I discover when I stumble along the dam on snowshoes. Tough-stemmed dry grasses have traced arcs in the snow where the wind has repeatedly blown them back and forth. In exposed areas, the sastrugi are as sharp-edged as though chiseled with a fine tool. Sastrugi is a lovely word, a Russian term for the ridges and furrows that wind planes into compacted snow.

The beaver lodge newly plastered with mud this autumn now lies covered in snow, as does the surrounding frozen pond surface. Deer tracks cross the pond, but otherwise there is no sign of life or activity. I imagine the beavers, warm and safe within the doubly insulated lodge, but also hungry perhaps. Winter is the lean time for them, and I hope that they cached enough food last summer.

The history of animal movement as recorded in snow.

Winter seals the beaver pond. A color version of this figure is included in the insert section.

Back up in the more sheltered pine groves on the terrace bordering the beaver meadow, I hear chickadees. A moment's pause reveals the little birds moving rapidly among the branches swaying in the wind. I look down to the meadow, where the air is thick and glittering with snow crystals entrained by the wind.

Taking the Pulse of the North St. Vrain
Beaver Meadow

After a year of observing this beaver meadow along North St. Vrain Creek and reading the observations and inferences of other naturalists and scientists, I have a better appreciation for Glynnis Hood's summary, in her book *The Beaver Manifesto*, of beavers as shapers of the physical and ecological landscapes of North America.

We can start with some facts and figures. Bob Naiman and Jerry Melillo studied beavers in portions of Quebec where the animals were largely unmolested by humans. In that low-relief landscape, the beavers influenced 30% to 50% of the total length of moderately sized rivers that ranged from about 3 feet to 45 feet wide. In a 115-square-mile area of the Kabetogama Peninsula of Minnesota, beaver ponds and meadows expanded from 1% of the landscape to 13% once the animals were no longer hunted or trapped by humans. In Rocky Mountain National Park, I estimate that beavers once occupied close to 1% of the park's total area. Today, beavers occupy about a tenth of what they used to occupy.

The numbers of species of all types of organisms increase because of the nutrient storage and diverse habitat created by the beavers. Plankton productivity increases in the warm, still water of ponds. Sediment deposited behind beaver dams creates ideal germination sites for willow, cottonwood, and alder. As accumulating sediment transforms the pond to a marsh, rushes, sedges, and other aquatic plants colonize the site. A beaver lodge is also more like an apartment complex than a single-family dwelling: scientists have documented muskrats, voles, mice, and even moths sharing a lodge with the beavers.

Ponds create habitat for aquatic invertebrates that prefer still water, including some species of insects, crayfish, and mussels. Terrestrial invertebrates as diverse as fruit flies, pine weevils, and leaf beetles are affected by the proportion of living to dead trees in a forest, and these proportions are influenced by the cutting and flooding activities of beavers. Amphibians such as salamanders and frogs favor beaver ponds as habitat. Turtles and water snakes like the ponds. Terrestrial snakes hunt along the water's edge. Deep ponds that do not freeze provide overwinter habitat for fish, as well as buffering seasonal fluctuations in water temperature. Ponds can act as source populations for fish, providing rich habitat that sustains large numbers of fish that migrate to adjacent areas from the pond. Waterfowl use the ponds for feeding, nesting, and rearing their young. Standing dead trees around the pond create cavity nesting sites for woodpeckers, other birds, and bats, as well as important perching sites for ospreys, hawks, owls, and eagles. The increased length of shoreline around the beaver-ponded water is hunted by kingfishers, mergansers, grebes, egrets, bitterns, and herons. Flycatchers, chickadees, and warblers feast on the insects hatching from the beaver ponds. Marshes developing from old ponds support snipe, woodcocks, partridges, turkeys, blackbirds, and sandpipers. Moose eat the aquatic vegetation and woody plants in and around the ponds. Elk and deer are attracted to the increased forage around the ponds. River otter, mink, water voles, and muskrat live in the ponds. Raccoons feed on invertebrates and fish in the ponds. Voles and shrews live in the meadows formed over old ponds, and predators come to eat the voles and shrews. Abandoned beaver lodges can be used by bobcats, pine marten, badgers, and red foxes.

This long list reflects the fact that riverside habitat is disproportionately important for a broad array of plants and animals relative to the fairly small proportion of any watershed that this habitat actually occupies. Even characteristically upland species such as sage grouse (*Centrocercus urophasianus* and *C. minimus*) depend on the herbaceous plants that grow along rivers for foraging late in the growing season. The frightening declines of sage grouse populations in the western United States—estimated at a 17% to 47% decline in various states since 1985—reflect several factors, but loss of suitable riverside habitat is one of them.

If I only visited the North St. Vrain meadow briefly, for perhaps an hour or two, I might find it hard to credit the abundance credited to beaver meadows. Days spent in the meadow throughout the seasons, however, convince me that life does indeed abound in this meadow where the overflowing creek acts like a watery magnet drawing in animals from the surrounding uplands. Much of the activity is cryptic, either because the organisms are microbes or invertebrates too small to observe readily or because the animals choose hours when I am not observing.

My graduate student Dan Scott set up a time-lapse camera aimed at a logjam farther up North St. Vrain Creek. Our intent was to record when, or if, the jam broke apart during peak flow. Two of the logs in the jam form a kinked bridge that spans the 60-foot-wide channel. We discovered that the jam provides a stream crossing for a wide variety of animals, most of which were photographed at night. We were delighted to see the infrared photos of ghostly white animals including bear, coyote, fox, bobcat, mountain lion, and pine marten walking the protruding logs like a tightrope. Seeing the photographs, one of my colleagues concluded that he would never set up a tent near a logjam. I concluded the opposite.

The diversity of life in the beaver meadow, so much greater than the adjacent uplands, reminds me of the tropics. In summer the beaver meadow is brightly green and wet like the tropics, too, even if the air is substantially cooler and drier than anything I've ever experienced at low latitudes. I think of the historical name for the Galapagos, Las Islas Encantadas, the Enchanted Isles, and I name the beaver meadow El Prado Encantado, the Enchanted Meadow.

All of the standing water, in ponds young and deep or shallowing on the way to becoming a marsh, buffers the valley bottom against disturbance. Drought and fire are less likely to stress the plants and animals of the wet valley bottom than species in the dry uplands. Even the record 2013 flood spread out and spent its energy surging around the dense willow stems and through the quiet ponds. All of these benefits result from the continued presence of beavers in the North St. Vrain beaver meadow.

Ecosystem Services

The abandoned beaver meadows in Rocky Mountain National Park represent much more widespread effects of loss of beaver populations. Ecologist Bob Naiman estimates that somewhere between 75,000 and 100,000 square miles of wetlands in the United States, much of which was likely beaver habitat, have been converted to agriculture or other uses since 1834. Historically, beaver activity resulted in storing large amounts of carbon, nitrogen, and other nutrients in the headwaters of watersheds. As beaver populations declined, these nutrients were instead transported downstream, sometimes with nasty consequences.

Starting in the 1970s, people began to describe ecosystem services. These are the vital functions that plants, animals, and ecosystems such as rivers provide for all living creatures, although humans too often ignore the services until they cease. Examples include the flood control that functioning floodplains provide or the clean water that results from physically complex river corridors in which water moves between the channel, floodplain, and underlying sediment, while microbes remove contaminants such as pathogens and excess nitrogen.

Nitrogen is a vital nutrient for living organisms, but diverse human activities, including combustion of fossil fuels and use of synthetic agricultural fertilizers, now completely dominate the global nitrogen cycle, introducing large amounts of nitrogen into the atmosphere and into river corridors in the form of nitrates. Nitrates can be broken down by microbes, but if excessive amounts of nitrates are present, even healthy microbial communities cannot keep up. The rivers of scenic Rocky Mountain National Park seem pristine when you labor up a long, steep trail to a high-elevation lake such as Thunder Lake upstream from the beaver meadow. But these rivers are in fact polluted with nitrates descending from the atmosphere with each rainfall, snowfall, or windy day. The nitrates originate in the farm fields, feedlots, and tailpipes of the more than four million people living along the base of the Colorado Front Range east of the national park. Upslope winds carry the nitrates aloft and then drop them onto the park's uplands and lakes, creating progressive acidification and changes in plant communities toward more nitrogen-tolerant species such as grasses and clovers. My colleague Jill Baron of the U.S. Geological Survey has documented these changes occurring over more than 20 years in the Loch Vale watershed just to the north of North St. Vrain Creek. The nitrates deposited from the atmosphere into high-elevation soils and waters then descend once more toward the base of the mountains in the water flowing through rivers that provides drinking water for the four million, creating the potential for algal blooms, blue baby syndrome, and non-Hodgkin's lymphoma.

Organic carbon is another vital building block of the tissues and respiratory function of most living organisms. Dissolved organic carbon in the drinking water hurts no one, but it discolors the water slightly and creates taste and odor. The standard response is to treat the water with chlorine, which creates carcinogenic trihalomethanes as a byproduct that is not removed from the drinking water. This is another of the many reasons to appreciate and restore the ecosystem services— including nitrogen and carbon storage—provided by active, functioning beaver meadows.

The case of methylmercury provides an interesting, if highly complex, illustration of the ecosystem effects of beaver meadows. Mercury is a naturally occurring element that human activities have dispersed throughout the environment. Ever hear of the phrase "mad as a hatter," or see one of the movie versions of *Alice in Wonderland* that features a mad hatter? Hatters in Europe had a much higher rate of insanity than the population as a whole during the 18th and 19th centuries because the traditional process used to make felt for hats involved mercury. Inorganic mercury was used to treat the fur of the small animals that went into felt, and the felt released mercury vapors that caused neurological damage in workers exposed to the vapors. Mercury is also used in placer gold mining, dispersing downstream from the mining site and through the bodies of organisms. And mercury is released through burning coal, alkali and metal processing, and medical and other wastes. Once in the atmosphere, mercury is widely disseminated and can circulate for years.

Mercury is poison with a capital P. The various mercury compounds have no known biological function, but they do interfere with most biological processes. Mercury is a carcinogen (it causes cancer), a mutagen (it causes chromosomal changes), and a teratogen (it causes developmental changes and abnormalities). Mercury bioaccumulates, concentrating within certain tissues of an organism as the organism consumes more and more mercury-contaminated food. And mercury biomagnifies, with organisms at higher levels of a food web receiving larger doses of mercury with each prey animal that they consume. Finally, mercury can travel adsorbed, or physically attached, to silt and clay particles, allowing the mercury to concentrate in river channels and floodplains. Yet we continue to use this highly toxic and dangerous compound in everything from thermometers to compact fluorescent light bulbs, thereby dispersing it throughout global environments. Elevated levels of mercury within organisms in mercury-contaminated areas can persist for at least a century after the source of pollution stops.

Regional and national assessments indicate high concentrations of mercury throughout soils, water, and living organisms across large parts of the Northern Hemisphere. The U.S. Geological Survey's National Water Quality Assessment program, for example, found mercury in most of the streambed sediments, invertebrates, and fish sampled across the United States during the 1990s and the early 21st century. Forms of mercury with relatively low toxicity can be converted to forms with much higher toxicity through biological processes. Sulfate-reducing bacteria and other microbes present in newly formed wetlands can convert mercury into methylmercury, which is one of the most toxic forms of mercury and is, unfortunately, readily taken up by other organisms. When beavers convert a flowing reach of river into a pond and flooded meadows, they promote methylation of mercury. This is only an issue, however, in newly flooded areas that have never before been flooded. In much less than a decade, methylation in new beaver ponds declines to levels comparable to other natural wetlands, and where beavers reoccupy historical beaver meadows, the difference in mercury methylation compared to other wetlands is negligible. Oded Levanoni and a group of scientists studying the effects of beaver ponds on methylmercury concentrations in Swedish streams thus concluded that how beaver activities affect the distribution and forms of mercury varies between sites depending on the age and history of beaver-influenced inundation of valley bottoms.

Betting on Beavers

Analogous to mercury, beaver engineering can both increase carbon storage in sediment and increase carbon emissions to the atmosphere in the form of methane. The relative importance of increased methane emissions enhancing greenhouse warming versus storage of nitrogen and organic carbon in the sediments of beaver meadows awaits a study that quantifies both effects, but I'm putting my money on the storage function and on the net positive environmental effects of beaver meadows. Over the course of the journey through this year, I've described how beaver meadows store water, sediment, and nutrients and how this storage creates a rich and diverse habitat that supports abundant organisms, from microbes to willows and salamanders to moose. Now we can step back and think about the cumulative effects of the historical loss of hundreds of thousands of beaver meadows across the Northern Hemisphere.

As the beavers and their dams disappeared, the meadows dried, the rivers cut down, and the wetland plants and animals disappeared. The rivers became leaky, passing everything from the uplands quickly downstream rather than retaining and processing it, converting nitrogen into plant tissue or carbon into mayflies and trout. The loss of wet valley bottoms makes rivers and streams more vulnerable to wildfires, floods, and warming climate. A dry elk grassland can readily burn during a fire that is less likely to consume a wet beaver meadow. A flood that concentrates all of its energy in a single channel is more likely to rip away at the streambed and banks than a flood that spreads out and loses velocity as it flows down a floodplain densely covered in willow thickets. And the droughts that will become more frequent as climate warms are more likely to wick away the flow of a stream in which all of the water is at the surface rather than safely tucked into the porous sediments beneath the channel.

Now, as the year comes to a close, I stand in the weak December sunlight and contemplate this beaver meadow that lies in an enormous trough with Longs Peak at the head and, at the downstream end, a deep notch cut through the terminal glacial moraine by North St. Vrain Creek. Here generations of beavers have created a complex and productive environment that supports their own colonies and dozens of other species. They require of us only to be left alone and protected from the seemingly relentless greed of our burgeoning human population. Surely we can save more of our beaver-dammed river segments and thus more of our ecosystem services and healthy rivers.

We have placed enormous stresses on the two beaver species of the Northern Hemisphere, hunting them nearly to extinction and destroying much of their potentially suitable habitat along rivers. Yet there is hope. European beavers have rebounded from an estimated 1,200 individuals at the start of the 20th century to more than a million animals today. North American beavers are also increasing in number, and natural resources managers are once again realizing that beavers can be, in the words of Joe Wheaton, cheap and cheerful, as well as highly effective, partners in river restoration.

One of the most effective means of change is to make enough people aware of a problem and to make them care about it. Empathy is a particularly important human trait in this context. As I write these words, an important change is occurring in how chickens are cared for because a sufficient number of people became upset about the suffering of hens spending their lives in tiny cages crowded together. Now many of the major egg purchasers, from McDonald's to Walmart, have announced that they will buy only cage-free eggs. Similarly, merchants sell shade-grown coffee because buyers care about songbirds. A movement to avoid products containing palm oil is gradually gaining adherents because of consumer concern about clearcutting of Asian rainforests to create palm-oil plantations.

I remember being skeptical when I first read Edwin James's 1820 description of beavers felling a tree. Then I read Hope Ryden's book *Lily Pond* and understood just how closely the members of a beaver colony can work together. I wish that *Lily Pond* was required reading for every North American. One would have to be completely lacking in empathy to read Ryden's descriptions of beaver play, affection, fear, and death without being attracted to the animals. As she is gradually dying, Lily, the matriarch of the beaver clan, turns to Ryden for help in obtaining food. A bond of trust and, at least on Ryder's side, affection has developed between Lily and this human who has appeared along the margins of the beaver pond at unpredictable intervals for four years. I think that if more people thought of beavers as individuals with distinct personalities and existences, rather than as nuisance rodents, protection of and appreciation for beavers would make a huge leap forward. Enos Mills attributed a high level of human-like sensibility and awareness to beavers. Who is to say he is wrong?

BIBLIOGRAPHY

General

B.W. Baker and E.P. Hill. 2003. Beaver. In, Wild mammals of North America: biology, management, and conservation, 2nd ed. G.A. Feldhammer, B.C. Thompson, and J.A. Chapman, eds. The Johns Hopkins University Press, Baltimore, MD, pp. 288–310.

E. Collier. 1959. Three against the wilderness. E.P. Dutton and Co., Inc., New York.

A. Radclyffe Dugmore. 1914. The romance of the beaver. J.B. Lippincott Company, Philadelphia, PA.

G.A. Hood. 2011. The beaver manifesto. Rocky Mountain Books, Victoria, Canada.

E.A. Mills. 1913. In beaver world. Houghton Mifflin, Boston, MA.

L.H. Morgan. 1868. The American beaver and his works. J. B. Lippincott & Co., Philadelphia, PA.

C.A. Morrison. 2003. Running the river: poleboats, steamboats, and timber rafts on the Altamaha, Ocmulgee, Oconee and Ohoopee. Salt Marsh Press, St. Simons Island, GA.

D. Müller-Schwarze and L. Sun. 2003. The beaver: natural history of a wetlands engineer. Comstock Publishing Associates, Ithaca, NY.

R. Olson and W.A. Hubert. 1994. Beaver: water resources and riparian habitat manager. University of Wyoming, Laramie, 48 pp.

Grey Owl. (Archibald Belaney). 1935. Pilgrims of the wild. Charles Scribner's Sons, New York.

H. Ryden. 1989. Lily Pond: four years with a family of beavers. William Morrow & Co., New York.

The Beaver Meadow on North St. Vrain Creek

C.B. Anderson, C.R. Griffith, A.D. Rosemond, R. Rozzi, and O. Dollenz. 2006. The effects of invasive North American beavers on riparian plant communities in Cape Horn, Chile: do exotic beavers engineer differently in sub-Antarctic ecosystems? Biological Conservation, vol. 128, pp. 467–474.

M.M. Cowan. 2003. Timberrr . . . a history of logging in New England. The Millbrook Press, Brookfield, CT.

R.J. Naiman, H. Decamps, and M. Pollock. 1993. The role of riparian corridors in maintaining regional biodiversity. Ecological Applications, vol. 3, pp. 209–212.

P.S. Ogden. 1961. Peter Skene Ogden's Snake Country journal, 1826–1827, vol. 23. K.G. Davies, ed. Hudson's Bay Record Society, London.

M. Oliver. 2016. Upstream: Selected essays. Penguin Press, New York.

M. Reuss. 2004. Designing the Bayous: The control of water in the Atchafalaya Basin, 1800–1995. Texas A&M University Press, College Station, TX.

J.R. Sedell and K.J. Luchessa. 1981. Using the historical record as an aid to salmonid habitat enhancement. In, Acquisition and utilization of aquatic habitat inventory information. N.B.

Armantrout, ed. American Fisheries Society Symposium Proceedings, Bethesda, MD, pp. 210–223.

E. Steele. 1841. A summer journey in the West. John Taylor, New York.

E.E. Wohl. 2001. Virtual rivers: lessons from the mountain rivers of the Colorado Front Range. Yale University Press, New Haven, CT.

E. Wohl. 2014. A legacy of absence: wood removal in U.S. rivers. Progress in Physical Geography, vol. 38, pp. 637–663.

January

William Bright. 2004. Colorado place names, 3rd ed. Johnson Books, Boulder, CO.

R. Olson and W.A. Hubert. 1994. Beaver: water resources and riparian habitat manager. University of Wyoming, Laramie.

F.M. Packard. 1947. A survey of the beaver population of Rocky Mountain National Park, Colorado. Journal of Mammalogy, vol. 28, pp. 219–227.

A.H. Zwinger and B.E. Willard. 1996. Land above the trees: a guide to the American alpine tundra, 4th ed. Johnson Books, Boulder, CO.

February

C.B. Anderson, C.R. Griffith, A.D. Rosemond, R. Rozzi, and O. Dollenz. 2006. The effects of invasive North American beavers on riparian plant communities in Cape Horn, Chile: do exotic beavers engineer differently in sub-Antarctic ecosystems? Biological Conservation, vol. 128, pp. 467–474.

B.W. Baker and E.P. Hill. 2003. Beaver. In, Wild mammals of North America: biology, management, and conservation, 2nd ed. G.A. Feldhammer, B.C. Thompson, and J.A. Chapman, ed. The Johns Hopkins University Press, Baltimore, Maryland, pp. 288–310.

W.W. Dalquest, F.B. Stangl, and M.J. Kocurko. 1990. Zoogeographic implications of Holocene mammal remains from ancient beaver ponds in Oklahoma and New Mexico. The Southwestern Naturalist, vol. 35, pp. 105–110.

M.A. Davis, M.K. Chew, R.J. Hobbs, A.E. Lugo, J.J. Ewel, and others. 2011. Don't judge species on their origins. Nature, vol. 474, pp. 153–154.

S. DeStefano, K.K.G. Koenen, C.M. Henner, and J. Strules. 2006. Transition to independence by subadult beavers (Castor canadensis) in an unexploited, exponentially growing population. Journal of Zoology, vol. 269, pp. 434–441.

C.A. Kaye. 1962. Early postglacial beavers in southeastern New England. Science, vol. 138, pp. 906–907.

B. Rains. 1987. Holocene alluvial sediments and a radiocarbon-dated relict beaver dam, Whitemud Creek, Edmonton, Alberta. The Canadian Geographer, vol. 31, pp. 272–277.

A.L. Swinehart and R.L. Richards. 2001. Palaecology of a northeast Indiana wetland harboring remains of the Pleistocene giant beaver (Castoroides ohioensis). Proceedings of the Indiana Academy of Sciences, vol. 110, pp. 151–166.

K.R.R. Swinnen, N.K. Hughes, and H. Leirs. 2015. Beaver (Castor fiber) activity patterns in a predator-free landscape: what is keeping them in the dark? Mammalian Biology, vol. 80, pp. 477–483.

C.J. Westbrook, D.J. Cooper, and C.B. Anderson. 2017. Alteration of hydrogeomorphic processes by invasive beavers in southern South America. Science of the Total Environment, vol. 574, pp. 183–190.

March

D.A. Burns and J.J. McDonnell. 1998. Effects of a beaver pond on runoff processes: comparison of two headwater catchments. Journal of Hydrology, vol. 205, pp. 248–264.

D.R. Butler. 1989. The failure of beaver dams and resulting outburst flooding: a geomorphic hazard of the southeastern Piedmont. The Geographical Bulletin, vol. 31, pp. 29–38.

D.R. Butler. 2012. Characteristics of beaver ponds on deltas in a mountain environment. Earth Surface Processes and Landforms, vol. 37, pp. 876–882.

D.R. Butler and G.P. Malanson. 1995. Sedimentation rates and patterns in beaver ponds in a mountain environment. Geomorphology, vol. 13, pp. 255–269.

D.R. Butler and G.P. Malanson. 2005. The geomorphic influences of beaver dams and failures of beaver dams. Geomorphology, vol. 71, pp. 48–60.

K.C. Green and C.J. Westbrook. 2009. Changes in riparian area structure, channel hydraulics, and sediment yield following loss of beaver dams. BC Journal of Ecosystems and Management, vol. 10, pp. 68–79.

G.A. Hood and S.E. Bayley. 2008. Beaver (Castor canadensis) mitigate the effects of climate on the area of open water in boreal wetlands in western Canada. Biological Conservation, vol. 141, pp. 556–567.

S. John and A. Klein. 2004. Hydrogeomorphic effects of beaver dams on floodplain morphology: avulsion processes and sediment fluxes in upland valley floors (Spessart, Germany). Quaternaire, vol. 15, pp. 219–231.

C.A. Johnston. 2012. Beaver wetlands. In, Wetland habitats of North America: ecology and conservation concepts. D.P. Batzer and A.H. Baldwin, eds. University of California Press, Berkeley, pp. 161–171.

R. Levine and G. Meyer. 2014. Beaver dams and channel sediment dynamics on Odell Creek, Centennial Valley, Montana, USA. Geomorphology, vol. 205, pp. 51–64.

R.K. Meentemeyer and D.R. Butler. 1999. Hydrogeomorphic effects of beaver dams in Glacier National Park, Montana. Physical Geography, vol. 20, pp. 436–446.

A. Morrison, C.J. Westbrook, and A. Bedard-Haughn. 2015. Distribution of Canadian Rocky Mountain wetlands impacted by beaver. Wetlands, vol. 35, pp. 95–104.

J. Nyssen, J. Pontzeele, and P. Billi. 2011. Effect of beaver dams on the hydrology of small mountain streams: example from the Chevral in the Ourthe Orientale basin, Ardennes, Belgium. Journal of Hydrology, vol. 402, pp. 92–102.

L. Persico and G. Meyer. 2012. Holocene beaver damming, fluvial geomorphology, and climate in Yellowstone National Park, Wyoming. Quaternary Research, vol. 71, pp. 340–353.

L. Persico and G. Meyer. 2013. Natural and historical variability in fluvial processes, beaver activity, and climate in the Greater Yellowstone Ecosystem. Earth Surface Processes and Landforms, vol. 38, pp. 728–750.

M.M. Pollock, T.J. Beechie, and C.E. Jordan. 2007. Geomorphic changes upstream of beaver dams in Bridge Creek, an incised stream channel in the interior Columbia River basin, eastern Oregon. Earth Surface Processes and Landforms, vol. 32, pp. 1174–1185.

M.M. Pollock, M. Heim, and D. Werner. 2003. Hydrologic and geomorphic effects of beaver dams and their influence on fishes. In, The ecology and management of wood in world rivers. S.V. Gregory, K. Boyer, and A. Gurnell, ed. American Fisheries Society, Bethesda, MD, pp. 213–233.

L.E. Polvi and E. Wohl. 2012. The beaver meadow complex revisited—the role of beavers in post-glacial floodplain development. Earth Surface Processes and Landforms, vol. 37, pp. 332–346.

L.E. Polvi and E. Wohl. 2013. Biotic drivers of stream planform: implications for understanding the past and restoring the future. BioScience, vol. 63, pp. 439–452.

C.J. Westbrook, D.J. Cooper, and B.W. Baker. 2006. Beaver dams and overbank floods influence groundwater–surface water interactions of a Rocky Mountain riparian area. Water Resources Research, vol. 42, pp. 1–12.

C.J. Westbrook, D.J. Cooper, and D.R. Butler. 2013. Beaver hydrology and geomorphology. In, Treatise on geomorphology. J. Shroder, ed. Ecogeomorphology, vol. 12, D.R. Butler and C.R. Hupp, eds. Academic Press, San Diego, pp. 293–303.

D.S. White. 1990. Biological relationships to convective flow patterns within stream beds. Hydrobiologia, vol. 196, pp. 149–158.

M.-K. Woo and J.M. Waddington. 1990. Effects of beaver dams on subarctic wetland hydrology. Arctic, vol. 43, pp. 223–230.

April

D.S. Brayton. 1984. The beaver and the stream. Journal of Soil and Water Conservation, vol. 39, pp. 108–109.

D. Burchsted and M.D. Daniels. 2014. Classification of the alterations of beaver dams to headwater streams in northeastern Connecticut, USA. Geomorphology, vol. 205, pp. 36–50.

D. Burchsted, M.D. Daniels, R. Thorson, and J. Vokoun. 2010. The river discontinuum: applying beaver modifications to baseline conditions for restoration of forested headwaters. BioScience, vol. 60, pp. 908–922.

D.R. Butler and G.P. Malanson. 1995. Sedimentation rates and patterns in beaver ponds in a mountain environment. Geomorphology, vol. 13, pp. 255–269.

M. De Visscher, J. Nyssen, J. Pontzeele, P. Billi, and A. Frankl. 2014. Spatio-temporal sedimentation patterns in beaver ponds along the Chevral River, Ardennes, Belgium. Hydrological Processes, vol. 28, pp. 1602–1615.

R.L. Ives. 1942. The beaver-meadow complex. Journal of Geomorphology, vol. 5, pp. 191–203.

C.A. Johnston. 2012. Beaver wetlands. In, Wetland habitats of North America: ecology and conservation concepts. D.P. Batzer and A.H. Baldwin, eds. University of California Press, Berkeley, pp. 161–171.

N. Kramer, E.E. Wohl, and D.L. Harry. 2012. Using ground penetrating radar to "unearth" buried beaver dams. Geology, vol. 40, pp. 43–46.

W.W. Macfarlane, J.M. Wheaton, N. Bouwes, M.L. Jensen, J.T. Gilbert, N. Hough-Snee, and J.A. Shivik. 2017. Modeling the capacity of riverscapes to support beaver dams. Geomorphology, vol. 277, pp. 72–99.

R.K. Meentemeyer and D.R. Butler. 1999. Hydrogeomorphic effects of beaver dams in Glacier National Park, Montana. Physical Geography, vol. 20, pp. 436–446.

L. Persico and G. Meyer. 2012. Holocene beaver damming, fluvial geomorphology, and climate in Yellowstone National Park, Wyoming. Quaternary Research, vol. 71, pp. 340–353.

M.M. Pollock, M. Heim, and D. Werner. 2003. Hydrologic and geomorphic effects of beaver dams and their influence on fishes. In, The ecology and management of wood in world rivers. S.V. Gregory, K. Boyer, and A. Gurnell, eds. American Fisheries Society, Bethesda, MD, pp. 23–233.

R. Ruedemann and W.J. Schoonmaker. 1938. Beaver-dams as geologic agents. Science, vol. 88, pp. 523–525.

C.J. Westbrook, D.J. Cooper, and B.W. Baker. 2011. Beaver assisted river valley formation. River Research and Applications, vol. 27, pp. 257–256.

May

B.W. Baker and E.P. Hill. 2003. Beaver. In, Wild mammals of North America: biology, management, and conservation, 2nd ed. G.A. Feldhammer, B.C. Thompson, and J.A. Chapman, eds. The Johns Hopkins University Press, Baltimore, MD, pp. 288–310.

R.A. Bartel, N.M. Haddad, and J.P. Wright. 2010. Ecosystem engineers maintain a rare species of butterfly and increase plant diversity. Oikos, vol. 119, pp. 883–890.

R.L. Beschta and W.J. Ripple. 2012. The role of large predators in maintaining riparian plant communities and river morphology. Geomorphology, vol. 157–158, pp. 88–98.

M.A. Briggs, L.K. Lautz, J.M. McKenzie, R.P. Gordon, and D.K. Hare. 2012. Using high-resolution distributed temperature sensing to quantify spatial and temporal variability in vertical hyporheic flux. Water Resources Research, vol. 48, 16 pp.

D.L. Correll, T.E. Jordan, and D.E. Weller. 2000. Beaver pond biogeochemical effects in the Maryland coastal plain. Biogeochemistry, vol. 49, pp. 217–239.

J.A. Crawford, R.A. Olson, N.E. West, J.C. Mosley, M.A. Schroeder, T.D. Whitson, R.F. Miller, M.A. Gregg, and C.S. Boyd. 2004. Ecology and management of sage-grouse and sage-grouse habitat. Journal of Range Management, vol. 57, pp. 2–19.

A.M. Gurnell. 1998. The hydrogeomorphological effects of beaver dam-building activity. Progress in Physical Geography, vol. 22, pp. 167–189.

G.A. Hood and S.E. Bayley. 2008. Beaver (Castor canadensis) mitigate the effects of climate on the area of open water in boreal wetlands in western Canada. Biological Conservation, vol. 141, pp. 556–567.

G.A. Hood and D.G. Larson. 2014. Beaver-created habitat heterogeneity influences aquatic invertebrate assemblages in boreal Canada. Wetlands, vol. 34, pp. 19–29.

C.A. Johnston. 2012. Beaver wetlands. In, Wetland habitats of North America: ecology and conservation concepts. D.P. Batzer and A.H. Baldwin, eds. University of California Press, Berkeley, pp. 161–171.

C.A. Johnston and R.J. Naiman. 1990. The use of a geographic information system to analyze long-term landscape alteration by beaver. Landscape Ecology, vol. 4, pp. 5–19.

L.K. Lautz, D.I. Siegel, and R.L. Bauer. 2006. Impact of debris dams on hyporheic interaction along a semi-arid stream. Hydrological Processes, vol. 20, pp. 183–196.

J.G. Lazar, K. Addy, A.J. Gold, P.M. Groffman, R.A. McKinney, and D.O. Kellogg. 2015. Beaver ponds: resurgent nitrogen sinks for rural watersheds in the northeastern United States. Journal of Environmental Quality, vol. 44, pp. 1684–1693.

R.J. Naiman, C.A. Johnston, and J.C. Kelley. 1988. Alteration of North American streams by beaver. BioScience, vol. 38, pp. 753–762.

R.J. Naiman and J.M. Melillo. 1984. Nitrogen budget of a subarctic stream altered by beaver (Castor canadensis). Oecologia, vol. 62, pp. 150–155.

R.J. Naiman, J.M. Melillo, and J.E. Hobbie. 1986. Ecosystem alteration of boreal forest streams by beaver (Castor canadensis). Ecology, vol. 67, pp. 1254–1269.

R.J. Naiman, G. Pinay, C.A. Johnston, and J. Pastor. 1994. Beaver influences on the long-term biogeochemical characteristics of boreal forest drainage networks. Ecology, vol. 75, pp. 905–921.

F. Rosell, O. Bozser, P. Collen, and H. Parker. 2005. Ecological impact of beavers Castor fiber and Castor canadensis and their ability to modify ecosystems. Mammal Reviews, vol. 35, pp. 248–276.

R. Ruedemann and W.J. Schoonmaker. 1938. Beaver-dams as geologic agents. Science, vol. 88, pp. 523–525.

J. Terwilliger and J. Pastor. 1999. Small mammals, ectomycorrhizae, and conifer succession in beaver meadows. Oikos, vol. 85, pp. 83–94.

P. Wegener, T. Covino, and E. Wohl. 2017. Beaver-mediated lateral hydrologic connectivity, fluvial carbon and nutrient flux, and aquatic ecosystem metabolism. Water Resources Research, vol. 53, pp. 4606–4623.

C.E. Wells, D. Hodgkinson, and E. Huckerby. 2000. Evidence for the possible role of beaver (Castor fiber) in the prehistoric ontogenesis of a mire in northwest England, UK. The Holocene, vol. 10, pp. 503–508.

C.J. Westbrook, D.J. Cooper, and B.W. Baker. 2006. Beaver dams and overbank floods influence groundwater–surface water interactions of a Rocky Mountain riparian area. Water Resources Research, vol. 42, W06404, 12 pp.

C.J. Westbrook, D.J. Cooper, and B.W. Baker. 2011. Beaver assisted river valley formation. River Research and Applications, vol. 27, pp. 247–256.

C.J. Whitfield, H.M. Baulch, K.P. Chun, and C.J. Westbrook. 2015. Beaver-mediated methane emission: the effects of population growth in Eurasia and the Americas. Ambio, vol. 44, pp. 7–15.

E.C. Wolf, D.J. Cooper, and N.T. Hobbs. 2007. Hydrologic regime and herbivory stabilize an alternative state in Yellowstone National Park. Ecological Applications, vol. 17, pp. 1572–1587.

J.P. Wright, C.G. Jones, and A.S. Flecker. 2002. An ecosystem engineer, the beaver, increases species richness at the landscape scale. Oecologia, vol. 132, pp. 96–101.

June

R.A. Bartel, N.M. Haddad, and J.P. Wright. 2010. Ecosystem engineers maintain a rare species of butterfly and increase plant diversity. Oikos, vol. 119, pp. 883–890.

S.T. Brown and S. Fouty. 2011. Beaver wetlands. Lakeline (Spring), pp. 34–38.

T.E. Dahl and G.J. Allord. 1997. History of wetlands in the conterminous United States. U.S. Geological Survey Water Supply Paper 2425, Washington, D.C., pp. 19–26.

B.B. Hanberry, J.M. Kabrick, and H.S. He. 2015. Potential tree and carbon storage in a major historical floodplain forest with disrupted ecological function. Perspectives in Plant Ecology, Evolution and Systematics, vol. 17, pp. 17–23.

A.W. Herre. 1940. An early Illinois prairie. American Botanist, vol. 46, pp. 39–44.

D.L. Hey and N.S. Phillipi. 1994. Reinventing a flood control strategy. Wetlands Initiative.

A. Law, F. McLean, and N.J. Wilby. 2016. Habitat engineering by beaver benefits aquatic biodiversity and ecosystem processes in agricultural streams. Freshwater Biology, vol. 61, pp. 486–499.

M.M. Pollock, T.J. Beechie, and C.E. Jordan. 2007. Geomorphic changes upstream of beaver dams in Bridge Creek, an incised stream channel in the interior Columbia River basin, eastern Oregon. Earth Surface Processes and Landforms, vol. 32, pp. 1174–1185.

A.M. Ray, A.J. Rebertus, and H.L. Ray. 2001. Macrophyte succession in Minnesota beaver ponds. Canadian Journal of Botany, vol. 79, pp. 487–499.

F. Rosell, O. Bozser, P. Collen, and H. Parker. 2005. Ecological impacts of beavers *Castor fiber* and *Castor canadensis* and their ability to modify ecosystems. Mammal Reviews, vol. 35, pp. 248–276.

E. Steele. 1841. A summer journey in the West. John Taylor, New York. (p. 123)

A.J. Veraart, B.A. Nolet, F. Rosell, and P.P. de Vries. 2006. Simulated winter browsing may lead to induced susceptibility of willows to beavers in spring. Canadian Journal of Zoology, vol. 84, pp. 1733–1742.

A. Vileisis. 1997. Discovering the unknown landscape: a history of America's wetlands. Island Press, Washington, D.C.

J.P. Wright. 2009. Linking populations to landscapes: richness scenarios resulting from changes in the dynamics of an ecosystem engineer. Ecology, vol. 90, pp. 3418–3429.

J.P. Wright, A.S. Flecker, and C.G. Jones. 2003. Local vs. landscape controls on plant species richness in beaver meadows. Ecology, vol. 84, pp. 3162–3173.

J.P. Wright, C.G. Jones, and A.S. Flecker. 2002. An ecosystem engineer, the beaver, increases species richness at the landscape scale. Oecologia, vol. 132, pp. 96–101.

July

N.L. Anderson, C.A. Paszkowski, and G.A. Hood. 2015. Linking aquatic and terrestrial environments: can beaver canals serve as movement corridors for pond-breeding amphibians? Animal Conservation, vol. 18, pp. 287–294.

R.S. Arkle and D.S. Pilliod. 2015. Persistence at the distributional edges: Columbia spotted frog habitat in the arid Great Basin, USA. Ecology and Evolution, vol. 5, pp. 3704–3724.

C.V. Baxter, K.D. Fausch, M. Murakami, and P.L. Chapman. 2004. Fish invasion restructures stream and forest food webs by interrupting reciprocal prey subsidies. Ecology, vol. 85, pp. 2656–2663.

J.-N. Beisel, P. Usseglio-Polatera, and J.-C. Moreteau. 2000. The spatial heterogeneity of a river bottom: a key factor determining macroinvertebrate communities. Hydrobiologia, vol. 422/423, pp. 163–171.

J.R. Benjamin, F. Lepori, C.V. Baxter, and K.D. Fausch. 2013. Can replacement of native by nonnative trout alter stream-riparian food webs? Freshwater Biology, vol. 58, pp. 1694–1709.

P. Collen and R.J. Gibson. 2001. The general ecology of beavers (*Castor* spp.) as related to their influence on stream ecosystems and riparian habitats, and the subsequent effects on fish—a review. Reviews in Fish Biology and Fisheries, vol. 10, p. 439–461.

M.R. Fuller and B.L. Peckarsky. 2011. Ecosystem engineering by beavers affects mayfly life histories. Freshwater Biology, vol. 56, p. 969–979.

Å. Hägglund and G. Sjöberg. 1999. Effects of beaver dams on the fish fauna of forest streams. Forest Ecology and Management, vol. 115, pp. 259–266.

G.A. Hood and D.G. Larson. 2014. Beaver-created habitat heterogeneity influences aquatic invertebrate assemblages in boreal Canada. Wetlands, vol. 34, pp. 19–29.

B.R. Hossack, W.R. Gould, D.A. Patla, E. Muths, R. Daley, K. Legg, and P.S. Corn. 2015. Trends in Rocky Mountain amphibians and the role of beaver as keystone species. Biological Conservation, vol. 187, pp. 260–269.

P.S. Kemp, T.A. Worthington, T.E.L. Langford, A.R.J. Tree, and M.J. Gaywood. 2012. Qualitative and quantitative effects of reintroduced beavers on stream fish. Fish and Fisheries, vol. 13, pp. 158–181.

M. McCaffery and L. Eby. 2016. Beaver activity increases aquatic subsidies to terrestrial consumers. Freshwater Biology, vol. 61, pp. 518–532.

D.M. McDowell and R.J. Naiman. 1986. Structure and function of a benthic invertebrate stream community as influenced by beaver (*Castor canadensis*). Oecologia, vol. 68, pp. 481–489.

S.C. Mitchell and R.A. Cunjak. 2007. Stream flow, salmon and beaver dams: roles in the structuring of stream fish communities within an anadromous salmon dominated stream. Journal of Animal Ecology, vol. 76, pp. 1062–1074.

M.M. Pollock, M. Heim, and D. Werner. 2003. Hydrologic and geomorphic effects of beaver dams and their influence on fishes. In, The ecology and management of wood in world rivers. S.V. Gregory, K. Boyer, and A. Gurnell, ed. American Fisheries Society, Bethesda, MD, pp. 23–233.

V.D. Popescu and J.P. Gibbs. 2009. Interactions between climate, beaver activity, and pond occupancy by the cold-adapted mink frog in New York State, USA. Biological Conservation, vol. 142, pp. 2059–2068.

J.A. Pounds, M.R. Bustamante, L.A. Coloma, J.A. Consuegra, M.P.L. Fogden, P.N. Foster, E. La Marca, K.L. Masters, A. Merino-Viteri, R. Puschendorf, S.R. Ron, G.A. Sanchez-Azofeifa, C.J. Still, and B.E. Young. 2006. Widespread amphibian extinctions from epidemic disease driven by global warming. Nature, vol. 439, pp. 161–167.

P. Rolauffs, D. Hering, and S. Lohse. 2001. Composition, invertebrate community and productivity of a beaver dam in comparison to other stream habitat types. Hydrobiologia, vol. 459, pp. 201–212.

I.J. Schlosser and L.W. Kallemeyn. 2000. Spatial variations in fish assemblages across a beaver-influenced successional landscape. Ecology, vol. 81, pp. 1371–1382.

J.M. Smith and M.E. Mather. 2013. Beaver dams maintain fish biodiversity by increasing habitat heterogeneity throughout a low-gradient stream network. Freshwater Biology, vol. 58, pp. 1523–1538.

J.W. Snodgrass and G.K. Meffe. 1998. Influence of beavers on stream fish assemblages: effects of pond age and watershed position. Ecology, vol. 79, pp. 928–942.

N. Weber, N. Bouwes, M.M. Pollock, C. Volk, J.M. Wheaton, G. Wathen, J. Wirtz, and C.E. Jordan. 2017. Alteration of stream temperature by natural and artificial beaver dams. PLOS One, vol. 12, e0176313.

August

T.G. Andrews. 2015. Coyote Valley: deep history in the high Rockies. Harvard University Press, Cambridge, MA.

B.W. Baker. 2005. Efficacy of tail-mounted transmitters for beaver. Wildlife Society Bulletin, vol. 34, pp. 218–222.

C.W. Buchholtz. 1983. Rocky Mountain National Park: a history. Colorado Associated University Press.

E.J. Dolin. 2010. Fur, fortune, and empire: the epic history of the fur trade in America. W.W. Norton & Co., New York.

J.C. Frémont. 1845. Report of the exploring expedition to the Rocky Mountains in the year 1842, and to Oregon and North California in the years 1843–44. Gales and Seaton, Washington, D.C.

E. James. 1823. Account of an expedition from Pittsburgh to the Rocky Mountains, performed in the years 1819 and '20. 2 vols. Carey and Lea, Philadelphia, PA.

C. Martin. 1978. Keepers of the game: Indian–animal relationships and the fur trade. University of California Press, Berkeley.

D. Mitchell, J. Tjornehoj, and B.W. Baker. 1999. Beaver populations and possible limiting factors in Rocky Mountain National Park, 1999. Unpublished report to Rocky Mountain National Park.

M. Wheatley. 1997. Beaver (*Castor canadensis*) home range size and patterns of use in the taiga of southeastern Manitoba: I. Seasonal variation. Canadian Field-Naturalist, vol. 111, pp. 204–210.

September

B.W. Baker, H.C. Ducharme, D.C.S. Mitchell, T.R. Stanley, and H.R. Peinetti. 2005. Interaction of beaver and elk herbivory reduces standing crop of willow. Ecological Applications, vol. 15, pp. 110–118.

R.L. Beschta and W.J. Ripple. 2008. Wolves, trophic cascades, and rivers in the Olympic National Park, USA. Ecohydrology, vol. 1, pp. 118–130.

J.M. Coles and B.J. Orme. 1983. *Homo sapiens* or *Castor fiber*? Antiquity, vol. LVII, pp. 95–102.

E. Collier. 1959. Three against the wilderness. E.P. Dutton and Co., Inc., New York.

S. DeStefano, K.K.G. Koenen, C.M. Henner, and J. Strules. 2006. Transition to independence by subadult beavers (*Castor canadensis*) in an unexploited, exponentially growing population. Journal of Zoology, vol. 269, pp. 434–441.

G. Feeley-Harnik. 2014. Bodies, words, and works: Charles Darwin and Lewis Henry Morgan on human–animal relations. In, America's Darwin: Darwinian theory and U.S. literary culture. T. Gianquitto and L. Fisher, eds. University of Georgia Press, Athens, pp. 265–301.

M. Francis. 2004. The strange career of the Canadian beaver: anthropomorphic discourses and imperial history. Journal of Historical Sociology, vol. 17, pp. 209–239.

D. Halley, F. Rosell, and A. Saveljev. 2012. Population and distribution of Eurasian beaver (*Castor fiber*). Baltic Forests, vol. 18, pp. 168–175.

D.E. Kroes and C.R. Hupp. 2010. The effect of channelization on floodplain sediment deposition and subsidence along the Pocomoke River, Maryland. Journal of the American Water Resources Association, vol. 46, pp. 686–699.

M.C. McKinstry and S.H. Anderson. 1993. Attitudes of private- and public-land managers in Wyoming, USA, toward beaver. Environmental Management, vol. 23, pp. 95–101.

D. Müller-Schwarze and L. Sun. 2003. The beaver: natural history of a wetlands engineer. Comstock Publishing Associates, Ithaca, NY.

G.P. Nicholas. 2007. Prehistoric hunter-gatherers in wetland environments: theoretical issues, economic organization and resource management strategies. In, Wetland archeology and environments: regional issues, global perspectives. M.C. Lille and S. Ellis, eds. Oxbow Books, Barnsley, UK, pp. 46–62.

B.A. Nolet and F. Rosell. 1998. Comeback of the beaver *Castor fiber*: an overview of old and new conservation problems. Biological Conservation, vol. 83, pp. 165–173.

A. Outwater. 1996. Water: a natural history. Basic Books, New York.

K. Spence. 2016. Natural hydrologists. Natural History. Online at http://jour.umt.edu/crown/stories/2016/Katy.php.

October

T.G. Andrews. 2015. Coyote Valley: deep history in the high Rockies. Harvard University Press, Cambridge, MA.

B.W. Baker, H.C. Ducharme, D.C.S. Mitchell, T.R. Stanley, and H.R. Peinetti. 2005. Interaction of beaver and elk herbivory reduces standing crop of willow. Ecological Applications, vol. 15, pp. 110–118.

R.L. Beschta and W.J. Ripple. 2008. Wolves, trophic cascades, and rivers in the Olympic National Park, USA. Ecohydrology, vol. 1, pp. 118–130.

J.M. Coles and B.J. Orme. 1983. *Homo sapiens* or *Castor fiber*? Antiquity, vol. LVII, pp. 95–102.

G. Feeley-Harnik. 2014. Bodies, words, and works: Charles Darwin and Lewis Henry Morgan on human–animal relations. In, America's Darwin: Darwinian theory and U.S. literary culture. T. Gianquitto and L. Fisher, eds. University of Georgia Press, Athens, pp. 265–301.

M. Francis. 2004. The strange career of the Canadian beaver: anthropomorphic discourses and imperial history. Journal of Historical Sociology, vol. 17, pp. 209–239.

D.E. Kroes and C.R. Hupp. 2010. The effect of channelization on floodplain sediment deposition and subsidence along the Pocomoke River, Maryland. Journal of the American Water Resources Association, vol. 46, pp. 686–699.

A. Leopold. 1949. A Sand County almanac. Oxford University Press, Oxford, UK.

K.N. Marshall, N.T. Hobbs, and D.J. Cooper. 2013. Stream hydrology limits recovery of riparian ecosystems after wolf reintroduction. Proceedings of the Royal Society Series B, vol. 280, 20122977, 7 pp.

E.C. Wolf, D.J. Cooper, and N.T. Hobbs. 2007. Hydrologic regime and herbivory stabilize an alternative state in Yellowstone National Park. Ecological Applications, vol. 17, pp. 1572–1587.

November

S. Albert and T. Trimble. 2000. Beavers are partners in riparian restoration on the Zuni Indian Reservation. Ecological Restoration, vol. 18, p. 87–92.

J. Baldwin. 2015. Potential mitigation of and adaptation to climate-driven changes in California's highlands through increased beaver populations. California Fish and Game, vol. 101, pp. 218–240.

N. Bouwes, N. Weber, C.E. Jordan, W.C. Saunders, I.A. Tattam, C. Volk, J.M. Wheaton, and M.M. Pollock. 2016. Ecosystem experiment reveals benefits of natural and simulated beaver dams to a threatened population of steelhead (*Oncorhynchus mykiss*). Scientific Reports, vol. 6, pp. 1–12.

D.S. Brayton. 1984. The beaver and the stream. Journal of Soil and Water Conservation, vol. 39, pp. 108–109.

J. Castro, M. Pollock, C. Jordan, G. Lewallen, and K. Woodruff. 2015. The beaver restoration guidebook: working with beaver to restore streams, wetland, and floodplains, version 1.0. U.S. Fish and Wildlife Service, Portland, OR, 189 pp. Online at http://www.fws.gov/oregonfwo/ToolsForLandowners/RiverScience/Beaver.asp.

P. DeVries, K.L. Fetherston, A. Vitale, and S. Madsen. 2012. Emulating riverine landscape controls of beaver in stream restoration. Fisheries, vol. 37, pp. 246–255.

D. Giriat, E. Gorczyca, and M. Sobucki. 2016. Beaver ponds' impact on fluvial processes (Beskid Niski Mts., SE Poland). Science of the Total Environment, vol. 544, pp. 339–353.

B. Goldfarb. 2018. Beaver dams without beavers? Artificial logjams are a popular but controversial restoration tool. Science. Online at https://www.sciencemag.org/news/2018/06/beaver-dams-without-beavers-artificial-logjams-are-popular-controversial-restoration.

J.E. Grasse. 1951. Beaver ecology and management in the Rockies. Journal of Forestry, vol. 49, pp. 3–6.

G.E. Johnson and C. van Riper III. 2014. Effects of reintroduced beaver (*Castor canadensis*) on riparian bird community along the Upper San Pedro River, southeastern Arizona and northern Sonora, Mexico. U.S. Geological Survey Open-File Report 2014–1121, Reston, VA.

J.G. Lazar, K. Addy, A.J. Gold, P.M. Groffman, R.A. McKinney, and D.Q. Kellogg. 2015. Beaver ponds: resurgent nitrogen sinks for rural watersheds in the northeastern United States. Journal of Environmental Quality, vol. 44, pp. 1684–1693.

W.W. Macfarlane, J.M. Wheaton, N. Bouwes, M.L. Jensen, J.T. Gilbert, N. Hough-Snee, and J.A. Shivik. 2017. Modeling the capacity of riverscapes to support beaver dams. Geomorphology, vol. 277, pp. 72–99.

R.A. Marston. 1994. River entrenchment in small mountain valleys of the western USA: influence of beaver, grazing and clearcut logging. Revue de Geographie de Lyon, v. 69, pp. 11–15.

R. Olson and W.A. Hubert. 1994. Beaver: water resources and riparian habitat manager. Cooperative Extension Service, University of Wyoming, Laramie, 48 pp.

M.M. Pollock, T.J. Beechie, J.M. Wheaton, C.E. Jordan, N. Bouwes, N. Weber, and C. Volk. 2014. Using beaver dams to restore incised stream ecosystems. BioScience, v. 64, pp. 279–290.

P. Rolauffs, D. Hering, and S. Lohse. 2001. Composition, invertebrate community and productivity of a beaver dam in comparison to other stream habitat types. Hydrobiologia, vol. 459, p. 201–212.

F. Rosell, O. Bozser, P. Collen, and H. Parker. 2005. Ecological impact of beavers *Castor fiber* and *Castor canadensis* and their ability to modify ecosystems. Mammal Review, vol. 35, pp. 248–276.

P.M. Scheffer. 1938. The beaver as an upstream engineer. Soil Conservation, vol. 3, pp. 178–181.

C. Vörösmarty, D. Lettenmaier, C. Leveque, M. Meybeck, C. Pahl-Wostl, J. Alcamo, W. Cosgrove, H. Grassl, H. Hoff, P. Kabat, F. Lansigan, R. Lawford, and R. Naiman. 2004. Humans transforming the global water system. EOS, Transactions of the American Geophysical Union, vol. 85, pp. 509, 513–514.

K. Woodruff and M.M. Pollock. 2015. Chapter 5—Relocating beaver. The beaver restoration guidebook: working with beaver to restore streams, wetland, and floodplains, version 1.0. U.S. Fish and Wildlife Service, Portland, OR, pp. 61–81. Online at http://www.fws.gov/oregonfwo/ToolsForLandowners/RiverScience/Beaver.asp.

December

E. James. 1823. Account of an expedition from Pittsburgh to the Rocky Mountains, performed in the years 1819 and '20. 2 vols. Carey and Lea, Philadelphia, PA.

C.A. Johnston. 2014. Beaver pond effects on carbon storage in soils. Geoderma, vol. 213, pp. 371–378.

O. Levanoni, K. Bishop, B.G. Mckie, G. Hartman, K. Eklöf, and F. Ecke. 2015. Impact of beaver pond colonization history on methylmercury concentrations in surface water. Environmental Science and Technology, vol. 49, pp. 12679–12687.

R.J. Naiman and J.M. Melillo. 1984. Nitrogen budget of a subarctic stream altered by beaver (*Castor canadensis*). Oecologia, vol. 62, pp. 150–155.

R.J. Naiman, J.M. Melillo, and J.E. Hobbie. 1986. Ecosystem alteration of boreal forest streams by beaver (*Castor canadensis*). Ecology, vol. 67, pp. 1254–1269.

R.J. Naiman, G. Pinay, C.A. Johnston, and J. Pastor. 1994. Beaver influences on the long-term biogeochemical characteristics of boreal forest drainage networks. Ecology, vol. 75, pp. 905–921.

E.W. Teale. 1965. Wandering through winter: a naturalist's 20,000 mile journey through the North American winter. Dodd, Mead, and Company, New York.

P. Wegener, T. Covino, and E. Wohl. 2017. Beaver-mediated lateral hydrologic connectivity, fluvial nutrient flux, and ecosystem metabolism. Water Resources Research, vol. 53, 4606–4623.

C.J. Whitfield, H.M. Baulch, K.P. Chun, and C.J. Westbrook. 2015. Beaver-mediated methane emission: the effects of population growth in Eurasia and the Americas. Ambio, vol. 44, 7–15.

E. Wohl. 2004. Disconnected rivers: linking rivers to landscapes. Yale University Press, New Haven, CT.

E. Wohl. 2013. Landscape-scale carbon storage associated with beaver dams. Geophysical Research Letters, vol. 40, pp. 1–6.

INDEX